Thomas Wermelinger

Analysis of Mechanical Stress and Microstructure by Raman Microscopy

Thomas Wermelinger

Analysis of Mechanical Stress and Microstructure by Raman Microscopy

A Multi-Dimensional Investigation at Small Length-Scales using Confocal Raman Microscopy

Südwestdeutscher Verlag für Hochschulschriften

Impressum / Imprint
Bibliografische Information der Deutschen Nationalbibliothek: Die Deutsche Nationalbibliothek verzeichnet diese Publikation in der Deutschen Nationalbibliografie; detaillierte bibliografische Daten sind im Internet über http://dnb.d-nb.de abrufbar.
Alle in diesem Buch genannten Marken und Produktnamen unterliegen warenzeichen-, marken- oder patentrechtlichem Schutz bzw. sind Warenzeichen oder eingetragene Warenzeichen der jeweiligen Inhaber. Die Wiedergabe von Marken, Produktnamen, Gebrauchsnamen, Handelsnamen, Warenbezeichnungen u.s.w. in diesem Werk berechtigt auch ohne besondere Kennzeichnung nicht zu der Annahme, dass solche Namen im Sinne der Warenzeichen- und Markenschutzgesetzgebung als frei zu betrachten wären und daher von jedermann benutzt werden dürften.

Bibliographic information published by the Deutsche Nationalbibliothek: The Deutsche Nationalbibliothek lists this publication in the Deutsche Nationalbibliografie; detailed bibliographic data are available in the Internet at http://dnb.d-nb.de.
Any brand names and product names mentioned in this book are subject to trademark, brand or patent protection and are trademarks or registered trademarks of their respective holders. The use of brand names, product names, common names, trade names, product descriptions etc. even without a particular marking in this work is in no way to be construed to mean that such names may be regarded as unrestricted in respect of trademark and brand protection legislation and could thus be used by anyone.

Verlag / Publisher:
Südwestdeutscher Verlag für Hochschulschriften
ist ein Imprint der / is a trademark of
OmniScriptum GmbH & Co. KG
Heinrich-Böcking-Str. 6-8, 66121 Saarbrücken, Deutschland / Germany
Email: info@svh-verlag.de

Herstellung: siehe letzte Seite /
Printed at: see last page
ISBN: 978-3-8381-1614-3

Zugl. / Approved by: Zürich, ETH Zürich, Diss., 2009

Copyright © 2010 OmniScriptum GmbH & Co. KG
Alle Rechte vorbehalten. / All rights reserved. Saarbrücken 2010

„But I'm more than just a little curious…."

Perfect Circle

Danksagung (Acknowledgement)

Ich bedanke mich zuallererst bei Prof. Ralph Spolenak, dass er mir die Möglichkeit gab, eine Doktorarbeit am Labor für Nanometallurgie zu machen. Er liess mich meine eigenen Ideen verwirklichen und unterstützte mich gegebenenfalls mit wertvollen Ratschlägen. Das Labor wurde gegründet nur wenige Monate bevor ich meine Arbeit anfing. Es war interessant zu sehen, wie sich die Gruppe während meiner Doktorarbeit entwickelte und gleichzeitig ein Teil dieser Entwicklung zu sein.

Weiter danke ich allen, die zum Gelingen meiner Arbeit beigetragen haben. Namentlich möchte ich Cesare Borgia, Flavio Mornaghini, Stephan Frank, Noble Woo, Susi Köppl, Matteo Seita, Giancarlo Pigozzi, Christian Solenthaler, Marianne Dietiker und Sven Olliges vom Labor für Nanometallurgie sowie Kilian Wasmer und Adrien Bidiville von der EMPA Thun für ihre Arbeit und Mithilfe danken.

Ein Dank geht auch an Sebastian Friess und die Leute bei WITec für die technische Unterstützung, wann immer ich Probleme oder Fragen bezüglich des Raman Mikroskops hatte.

Ein Dankeschön möchte ich dem ganzen LNM Team und speziell Marianne Dietiker, Sven Olliges und Karen Magid machen, da sie der Grund waren (sind) für die gute Atmosphäre am Arbeitsplatz. Sven Olliges war bei jeglicher Art von Physikproblemen ein wichtiger Ansprechpartner. Marianne Dietiker machte sich speziell unentbehrlich, da sie sich unentwegt um die Ordnung auf meinem Tisch, das Überleben meiner Pflanzen und meinen Bartwuchs sowie die Innenarchitektur meiner Wohnung kümmerte. Sie versuchte mich zu einem besseren Menschen zu machen.

Glücklicherweise entschieden sich einige Studenten dafür bei mir ein Projekt zu machen. Namentlich waren das: Dario Corica, Julian Schneider, Tobias Frey, Claudia Müller,

Christophe Charpentier, Müge Yüksek, Enrico Bigi, Philipp Chen and Marta Szankowska. Ihre Hilfe, ihre Resultate und ihre Überlegungen erleichterten mir die Arbeit sehr.

Bedanken möchte ich mich auch ganz herzlich bei all meinen Freunden und Kollegen in- und ausserhalb der ETH. Sie waren eine grosse Unterstützung während dieser Zeit und haben meinen endlosen Geschichten über meine Arbeit an der ETH ohne zu nörgeln ertragen. Ich denke, den meisten ist bewusst, dass sich das in Zukunft kaum ändern wird.

Letztendlich danke ich meiner ganzen Familie für ihre Unterstützung während des gesamten Studiums als auch während der Doktorarbeit. Ohne ihre Hilfe wäre mein Studium und somit diese Arbeit nicht möglich gewesen.

Euch allen ein grosses DANKESCHÖN.

Summary

Goal of the thesis was to measure and characterize multi-dimensional mechanical stresses and changes of the microstructure in materials at small length-scales by means of Raman microscopy. The work is motivated by the constant progress and development in the fields of microelectronics and the resulting ongoing miniaturization of its components. Miniaturization can cause new kinds of stress related problems in materials which may have a negative influence on the reliability and durability of devices. Stresses and changes of the microstructure can also affect optical and electronic properties of semiconductors such as zinc oxide (ZnO). To fully understand the behavior and properties of materials on the nano- and micro-scopic scale, it is essential to measure stresses with the highest possible lateral resolution. At present, several techniques are available for analyzing stresses and changes of the microstructure with a high spatial resolution such as Raman microscopy, X-ray microdiffraction or convergent beam electron diffraction. All methods have their advantages and drawbacks. Raman microscopy was the method of choice as it features a spatial resolution in the submicrometer range, a high data acquisition rate as well as high strain sensitivity.

Experiments in one-, two- and three-dimensions were performed focusing each time on different aspects of mechanical stresses and changes of the microstructure. The findings of the work are published in five full length articles and can be divided into two main parts. The first part mainly covers the methodology of Raman microscopy itself. The second part concentrates on the measurements of stresses and properties of different materials.

Raman Microscopy

In a full length article about compression tests performed on 8 µm diameter silicon pillars combined with a micro-Raman set-up, we have shown that Raman microscopy is an accurate method for measuring mechanical stress. The uniaxial stress in the micro-pillars was applied with a micro-indenter while Raman measurements were performed *in situ*. The deviation between the stresses calculated from the Raman peak shifts and a load cell was less than 3% throughout the whole test until the rupture of the pillars.

In the third chapter of the thesis, a method for the mapping of 2-dimensional stresses in metal thin films was developed by means of Raman microscopy using a silicon thin film as a strain gage material. The sample consisted of layers of 280 nm poly-silicon, 30 nm silicon nitride and 850 nm aluminum. At 200°C the measured thermal stresses in the silicon film were on the order of +250 MPa to +450 MPa which directly corresponded to compressive stresses of -90 MPa to -160 MPa in the aluminum film.

This work presents in the chapters 5 and 6 illustrate a novel approach in measuring and visualizing three dimensional stresses in transparent, Raman active materials using confocal Raman microscopy. Two cases studies on a ZnO single crystal and a sapphire single crystal were performed with a spatial resolution in the sub micrometer range. Although, the method was more of a qualitative nature than a quantitative one it is a very useful tool for visualizing 3-D stress fields. Quantifying stresses requires the presumption of certain defined stress states.

Besides measuring stresses, the laser of the Raman microscope was used to induce grain growth in amorphous silicon thin films. The focused laser with a spot size of 300 nm was used to successfully produce well-defined, crystalline structures with a diameter in the range of one micrometer. This could be a useful tool to fabricate freestanding structures or electrical conductors.

Materials Aspects

Raman and TEM analysis showed that the main deformation mechanism of the silicon micro pillars is brittle fracture. No phase transformations and nucleation were observed. Also the movement of dislocations seems to play a minor role. The average compressive failure strength of the micro-pillars was -5.1 GPa. The authors have indications that Raman microscopy allowed observing cracks propagating through the pillars.

A lamella was extracted from the center of a micro indent in a silicon wafer by means of focused ion beam (FIB). 2-D Raman maps showed that the width of the Raman peaks correlates to the defect density of the lamella, which was analyzed by TEM. Further, the study demonstrated nicely how implanted Ga^+-ions, originating from the lamella extraction process, shift the Raman signal to lower wavenumbers. This averted the measurement of residual stresses in the lamella and leads to the question how much the sample preparation changes the properties of the sample itself.

The residual stress fields in the vicinity of micro indents in a sapphire single crystal and in a ZnO single crystal were analyzed and visualized in 3-D with a confocal Raman microscope. Two studies, presented as full length articles, showed that the residual stress field is mainly influenced by the deformation mechanism of the material and not by the indenter geometry. While the indents in both materials exhibited a fourfold symmetry, the residual stress field in sapphire and in ZnO possessed a threefold and a twofold symmetry, respectively. In the case of ZnO the biaxial stress field at the surface of the crystal was calculated. The stress field is highly anisotropic as the distortion caused by the dislocations is much stronger along the x-axis then along the c-axis. In both materials an evolution of the microstructure was observed. Sapphire exhibited a phase transformation directly below the indent while in ZnO the volume of the area with a high dislocation density (226 μm^3) could be calculated from the 3-D maps.

Zusammenfassung

Ziel der Arbeit war es mittels Raman Mikroskopie mechanische Spannungen und Veränderungen der Mikrostruktur in Materialien ein-, zwei- und dreidimensional zu messen. Die Motivation der Arbeit liegt hauptsächlich in den andauernden Fortschritten und Entwicklungen auf dem Gebiet der Mikroelektronik. Durch die ständige Verkleinerung der einzelnen Komponenten und Bauteile können in den verwendeten Materialien neue spannungsbedingte Probleme auftauchen, welche die Zuverlässigkeit und die Dauerhaftigkeit negativ beeinflussen können. Mechanische Spannungen und Veränderungen der Mikrostruktur beeinflussen unter Umständen auch die optischen und elektrischen Eigenschaften von Halbleitern wie zum Beispiel Zinkoxid (ZnO). Um das Verhalten und die Eigenschaften von Materialien im Nano- und Mikrometerbereich vollständig zu verstehen, ist es wichtig, dass man die mechanischen Spannungen als auch Veränderungen der Mikrostruktur mit einer grösstmöglichen örtlichen Auflösung messen kann. Mehrere verschiedene Methoden wie zum Beispiel Röntgen-Mikrobeugung, konvergente Elektronenstrahlbeugung und Raman Mikroskopie ermöglichen es heutzutage Spannungen mit einer hohen örtlichen Auflösung zu messen. Alle Methoden haben ihre Vor- und Nachteile und eigenen sich für eine bestimmte Art von Messungen. Raman Mikroskopie wurde für diese Arbeit gewählt, weil die Methode eine hohe örtliche Auflösung hat, schnell Daten akquiriert und sehr dehnungssensitiv ist.

Im Rahmen der Arbeit wurden ein-, zwei-, und dreidimensionale Untersuchungen durchgeführt, welche sich jeweils auf bestimmte Aspekte der Thematik von mechanischen Spannungen und Veränderung der Mikrostruktur konzentrierten. Die Ergebnisse der Arbeit wurden in fünf Artikeln veröffentlicht und können grundsätzlich in zwei Teile unterteilt werden. Ein erster Teil der Resultate bezieht sich vor allem auf die Methode der Raman Mikroskopie, während der zweite Teil auf materialspezifische Erkenntnisse fokussiert.

Raman Mikroskopie

Im ersten Kapitel, welches das Verhalten von 8 µm breiten Siliziumsäulen unter einachsigen Druckspannungen untersucht, konnte bewiesen werden, dass Raman Mikroskopie eine genaue Methode ist um Spannungen zu messen. Die gemessenen Spannungen mittels Raman Mikroskopie und mittels einer Kraftmesszelle wichen während der ganzen Messung bis zum Bruch der Säulen nie mehr als 3% voneinander ab.

In einem weiteren Kapitel wurde eine Methode entwickelt um mechanische Spannungen zweidimensional in dünnen Metallfilmen zu messen. Dabei wurde eine polykristalline Siliziumdünnschicht als Dehnmessstreifen benutzt. Der Mehrschichtenverbund bestand aus 280 nm poly-Silizium, 30 nm Siliziumnitrid und 850 nm Aluminium. Bei einer Temperatur von 200°C wurden in der Siliziumschicht Zugspannungen von +250 MPa bis +450 MPa gemessen was Druckspannungen von -90 MPa bis -160 MPa in der Aluminiumschicht entspricht.

Die Doktorarbeit präsentiert einen neuartigen Ansatz um mechanische Spannungen in transparenten, Raman aktiven Materialen zu messen und darzustellen. In zwei Fallbeispielen, durchgeführt in Saphir und Zinkoxid, konnten die Eigenspannungen rund um einen Indent dreidimensional gemessen werden. Die räumliche Auflösung war dabei besser als ein μm^3. Die Messungen haben allerdings eher einen qualitativen als einen quantitativen Charakter. Um quantitative Aussagen möglich zu machen, müssen bestimmte Annahmen bezüglich des herrschenden Spannungszustands gemacht werden.

Neben dem Messen von Spannungen war es auch möglich, den Laser des Raman Mikroskops zu verwenden, um Kornwachstum in einem amorphen Siliziumdünnfilm zu generieren. Da der fokussierte Laser einen Durchmesser von minimal 300 nm hatte, konnten kristalline Strukturen in der Grössenordnung von ungefähr einem Mikrometer erzeugt werden. Diese Methode könnte Anwendung beim Produzieren von freistehenden, kristallinen Strukturen und elektrischen Leiterbahnen finden.

Materialspezifische Aspekte

Analysen durchgeführt mit Raman Mikroskopie und TEM zeigten, dass der Bruchmechanismus von Siliziummikrosäulen hauptsächlich spröde ist. Es wurde keine Phasentransformationen beobachtet. Nukleierung von Versetzungen als auch die Bewegung von Versetzungen scheinen eher eine untergeordnete Rolle zu spielen. Raman Mikroskopie ermöglichte es, Risse zu detektieren, welche durch die Siliziummikrosäulen liefen.

Eine Lamelle wurde mittels Focused Ion Beam (FIB) aus dem Zentrum eines Mikroindents, platziert in einem Siliziumwafer, extrahiert. Zweidimensionale Raman Messungen, durchgeführt auf der Lamelle, zeigten dabei, dass die Breite des Silizium Ramansignals sehr gut mit der Defektdichte rund um den Indent korreliert. Weiter wurde ersichtlich, dass die Ga^+-Ionen, welche zum Extrahieren der Lamelle verwendet werden, in die Oberfläche implantiert werden. Die implantierten Ionen führen zu einer Veränderung des Ramansignals. Dies verunmöglichte die Messung von Eigenspannungen in der Lamelle.

Die Eigenspannungen, welche sich rund um Mikroindents bilden, konnten in Saphir als auch in ZnO dreidimensional gemessen werden. Beide Studien wurden als Artikel veröffentlicht und zeigten, dass die räumliche Anordnung von Eigenspannungen hauptsächlich von den Mechanismen der plastischen Deformationen des jeweiligen Materials abhängig ist und nicht von der Geometrie der verwendeten Indenterspitze. Im Falle von ZnO konnte das biaxiale Spannungsfeld an der Kristalloberfläche berechnet werden. Die Resultate weisen darauf hin, dass die Eigenspannungen stark anisotrop sind, da die Verzerrungen durch die Versetzungen entlang der x-Achse viel stärker sind als entlang der c-Achse. In beiden Materialien konnten Veränderungen der Mikrostruktur gemessen werden. Direkt unter dem Eindruck wurde in Saphir eine Phasentransformation beobachtet. In ZnO war es möglich, das Volumen des Gebietes mit einer hohen Defektdichte (226 μm^3) zu berechnen.

Table of Content

DANKSAGUNG (ACKNOWLEDGEMENT)..5

SUMMARY..7

    Raman Microscopy...8

    Materials Aspects...9

ZUSAMMENFASSUNG..10

    Raman Mikroskopie...11

    Materialspezifische Askpekte..12

TABLE OF CONTENT..13

1   INTRODUCTION 15

    1.1   Motivation and Aim of the Thesis 15

    1.2   Thesis Outline 18

    1.3   Short Historical Review of Raman Spectroscopy 19

    1.4   Fundamentals of Raman Scattering.. 20

            1.4.1   Measuring in Conventional Backscattering Configuration 24

            1.4.2   Off-axis Raman Spectroscopy... 27

            1.4.3   Stress Tensor Analysis in Backscattering Raman Microscopy 30

    1.5   References 34

2   *IN SITU* COMPRESSION TESTS ON MICRON SIZED SILICON PILLARS BY RAMAN MICROSCOPY-STRESS MEASUREMENTS AND DEFORMATION ANALYSIS 39

    2.1   Abstract 39

    2.2   Introduction 40

    2.3   Experiments and Materials... 41

    2.4   Results.. 44

    2.5   Discussion 49

2.6	Conclusions		52
2.7	Acknowledgements		53
2.8	Appendix		54
	2.8.1	The Influence of Stress on the Peak Position of Polymers	54
	2.8.2	Acknowledgement (Appendix)	58
2.9	References		59

3 MEASURING STRESSES IN METAL THIN FILMS BY MEANS OF RAMAN MICROSCOPY USING SILICON AS A STRAIN GAGE MATERIAL ... 63

3.1	Abstract		64
3.2	Introduction		64
3.3	Experimental		66
3.4	Results		69
	3.4.1	Silicon Thin Film	69
	3.4.2	Aluminum Thin Film	72
3.5	Discussion		73
	3.5.1	Grain Growth in Silicon Thin Films	73
	3.5.2	Grain Size	74
	3.5.3	Stress Measurement	75
3.6	Conclusions		83
3.7	Acknowledgement		84
3.8	References		85

4 CORRELATING RAMAN PEAK SHIFTS WITH PHASE TRANSFORMATION AND DEFECT DENSITIES: A COMPREHENSIVE TEM AND RAMAN STUDY ON SILICON ... 91

4.1	Abstract		91
4.2	Introduction		92
4.3	Experimental		94
4.4	Results		96
4.5	Discussion		102
4.6	Conclusions		108
4.7	Acknowledgement		109
4.8	Appendix		109
	4.8.1	Synchrotron X-Ray Laue Microdiffraction on He^+-ion Implanted Silicon	109
	4.8.2	Si^{2+}-ion Implantation in Silicon	113
4.9	References		118

5 3-D RAMAN SPECTROSCOPY MEASUREMENTS OF THE SYMMETRY OF RESIDUAL STRESS FIELDS IN PLASTICALLY DEFORMED SAPPHIRE CRYSTALS 123

5.1 Abstract 123

5.2 Introduction 124
5.2.1 Raman Effect of Sapphire 125
5.2.2 Experimental 126

5.3 Results 128

5.4 Discussion 135
5.4.1 Mechanical Aspects of Indenting Sapphire 135
5.4.2 3-D Raman Microscopy 138

5.5 Summary and Conclusions 140

5.6 Acknowledgments 141

5.7 References 142

6 SYMMETRY OF RESIDUAL STRESS FIELDS OF ZNO BELOW AN INDENT MEASURED BY 3-D RAMAN SPECTROSCOPY 145

6.1 Abstract 145

6.2 Introduction 146

6.3 Experimental 148

6.4 Results 150

6.5 Discussion 154

6.6 Conclusion 160

6.7 Acknowledgement 161

6.8 References 162

7 CONCLUSION AND OUTLOOK 167

7.1 Conclusion 167
7.1.1 Raman Microscopy 167
7.1.2 Materials Aspects 168

7.2 Outlook 170

1 Introduction

1.1 Motivation and Aim of the Thesis

Microelectronics have become an essential part of our every day lives. Progress in the field of microelectronics is often equalized with miniaturization. The frequently quoted Moore's law states that the amounts of stored data per area will double every 18 month [1]. Although he stated that his prediction should be valid only until 1980 the developments in microelectronics are up to nowadays continuing this trend. One of the big tasks of ongoing research is the continuing reduction of the feature size. With miniaturization of devices stress related problems can occur which might have a negative influence on the reliability of the system. Electromigration for example is a well known failure mechanism in thin film conductors and microelectronics [2]. Korhonen *et al.* illustrated in their work [3] that electromigration can cause high mechanical stresses in interconnects which might trigger voids and failures. Stresses occur also during thin film deposition, for instance, the formation of chemical vapor deposited silicon nitride films. In this case, the occurring stresses can be adjusted by changing the deposition temperature as well as the stoichiometry of the gases [4]. Differences in the thermal expansion coefficient of different materials can induce thermal stresses if the temperature changes [5]. Another example is the fatigue in micron-sized poly- and single crystalline silicon structures [6]. Mechanical stresses can not only lead to failures but they can also change the electrical and optical properties of materials such as ZnO [7].

All these samples show how important it is to have reliable methods for measuring mechanical stresses with a high lateral resolution. Several methods are available nowadays but they all have their drawbacks. X-ray microdiffraction [8-10] allows measuring the strain tensor but is based on large scale facilities and measuring time is generally limited. Convergent beam electron diffraction with a transmission electron microscope (TEM) [11] has the possibility to measure stress with a lateral resolution of only a few nanometers, but the samples have to be thinned to become electron transparent. Therefore the stress state under investigation is usually relaxed or altered in comparison to the original stress state. Electron induced Kossel diffraction with a scanning electron microscope is another accurate method for measuring stresses (SEM) [12] but at the current state of the technique lacks automation. The strain measurements by means of Electron backscatter diffraction (EBSD) with an SEM [13] deduces underlying stress by observing the patterns of electrons scattered back from the crystal planes. However, the pattern quality depends critically on various factors such as the interaction volume (beam defocus, electron beam energy and current), the sample surface condition (oxide layer, contamination with hydrocarbons) and in some cases even the crystal orientation [14]. All of these methods have two aspects in common; they all measure elastic strain in a crystalline material and require the knowledge of its elastic properties to convert strain into stress. All diffraction methods, independent of whether photons or electrons are the probe utilized, directly provide a measure of the lattice parameter of crystalline materials. Raman spectroscopy on the other hand, is a spectroscopy technique which yields the energies of the phonon spectrum. In contrast to diffraction techniques which solely require crystal periodicity, Raman spectroscopy requires a material to be polarizable, which limits the materials choice to ceramics, polymers and semiconductors. Raman studies on metals [15] rely on the mechanical coupling of a Raman active material to the metal and subsequent mechanical modeling. While Raman spectroscopy has traditionally been employed for studies of stress in microelectronic materials [16-20], strain measurements by Raman microscopy and the conversion to mechanical stress is at present being extended to other materials classes and applications. In these cases, phonon deformation potentials still need to be established. Thus, let us investigate the advantages and draw-backs of this technique.

The advantages are:
- High lateral resolution
- Depth resolution
- Available in laboratories at moderate cost

- Fast and easy data acquisition
- High strain sensitivity

In contrast the disadvantages include:
- Only certain classes of materials are Raman active
- Phonon deformation potentials are only available for a small set of materials
- Materials need to be transparent for 3-D mapping
- Complete strain tensor is difficult to access
- Fluorescence may overshadow Raman peaks

The following table summarizes and compares the current technique for local strain measurements which are mostly based on diffraction.

Table 1.1 Comparison of the Raman technique for stress measurement to comparable electron and X-ray diffraction based techniques

	Raman	Electron diffraction		X-ray microdiffraction		
		EBSD	CBED	Laue	Monochromatic	Kossel
Lateral resolution	300 nm	20 nm	5 nm	500 nm	50 nm	500 nm
Depth resolution	500 nm	-	-	1 μm	-	-
Ease of use	Lab based	SEM	TEM	Synchrotron	Synchrotron	SEM
Strain resolution	10^{-4}	10^{-3}	10^{-4}	10^{-4}	10^{-4}	10^{-5}
Time per pixel	0.1 s	1 s	1 s	5 s	10 s	10 s
Materials	Polarizable	Single Crystalline	Electron transparent	Single crystalline	Crystalline	Single crystalline
Strain tensor components	1-3	2-5	2-5	5-6	1-2	6

The conversion of strain into stress by Hooke's law is common to all methods. Complex strain states cannot currently be accessed by Raman microscopy. This draw-back will be overcome in the near future. With regards to lateral resolution, Raman microscopy is on the lower side for the methods compared. Here, two possibilities are viable. First, near-field optics could be employed to achieve resolutions below 100 nm. The issue with this option is that both near-field optics and the Raman effect strongly attenuate the primary signal and if used in conjunction leave just a few photons of signal to count. The other option is to locally enhance the signal by Plasmon resonances, e.g. by putting metallic nanoparticles of gold or silver in a focused beam [21-23].

The aim of this work was to measure and analyze mechanical stresses and stress related effects such as microstructural changes on a very local level by means of confocal Raman microscopy. Depending on the experimental set-up, one, two or three dimensional measurements were performed with a lateral resolution in the submicrometer range.

1.2 Thesis Outline

The thesis begins with a historical review of the developments of the Raman spectroscopy and its possible applications. In the first chapter the fundamentals of Raman microscopy are explained including a section about the principles of strain measurements by Raman microscopy in the case of silicon. The main part consists of five publications which are either already published in or submitted to peer reviewed journals. Chapter 2 presents the results of one dimensional *in situ* compression tests on silicon micro pillars. The outcome shows the high accuracy of Raman spectroscopy for measuring stresses. In Chapter 3, a method is proposed to measure stresses in thin metallic films, therefore in 2-D, by Raman microscopy using silicon as a strain gage material. A two dimensional study comparing the microstructural details of a cross-section extracted from an indent in silicon and the Raman peak shape and peak width is discussed in Chapter 4. The Chapter 5 is dedicated to 3-D-Raman-Spectroscopy measurements of the symmetry of residual stress fields in plastically deformed sapphire crystals. The last chapter (Chapter 6) presents a detailed study of the stress field and the defect structure in 3-D around an indent on a ZnO single crystal. Finally, some general conclusions are drawn and suggestions for future work made.

1.3 Short Historical Review of Raman Spectroscopy

The Raman effect was discovered by the two Indian physicists C. V. Raman and K.S. Krishnan in 1928 [24]. Independently, the same effect was observed in 1928 by Landsberg and Mandelstam[25]. In 1930, Raman won the Noble Prize in Physics for his discovery. The first studies using the Raman effect were performed on gases and liquids [26-28]. The invention of the laser increased the possible applications of the Raman spectroscopy remarkably. For a long time Raman spectroscopy was mostly applied in chemical studies as a complementary technique to infrared spectroscopy, giving information on the chemical composition and crystallinity of the sample. In the late 1960's and early 1970's studies executed on diamond and silicon revealed the pressure dependence of the Raman signal of semiconductors and ceramics [17, 29, 30]. Another big evolutionary step in the history of Raman spectroscopy was the combination of the Raman set-up with light microscopy. It enabled to measure objects with a lateral size of few micrometers [31]. Brueck *et al.* and Lyon *et al.* were in 1982 the first who used Raman microscopy to measure stresses with a lateral resolution in the order of 1 to 5 µm [32, 33]. Since then Raman microscopy has become a powerful tool for analyzing stresses in semiconductors and ceramics which will be discussed in more details in the next sections. In 1990 confocal Raman microscopy was used to map living cells and in another case such a set-up was used in the same year to perform 3-D measurements [34, 35]. Another effect of Raman spectroscopy attracts a lot of attention. The Raman scattering in the vicinity of noble metal nanoparticles is drastically enhanced [36]. This so-called surface enhanced Raman scattering (SERS) effect even allows detecting single molecules [22]. More recent results showed that it is possible to use the SERS effect to significantly lower the lateral resolution down to 15 nm [37]. Raman spectroscopy has become in the last 80 years a powerful tool to answer different question in multiple scientific fields such as biology, chemistry, physics and material science and can be applied to macroscopic as well as nanoscopic devices.

1.4 Fundamentals of Raman Scattering

Raman scattering is used to obtain information about the structure and properties of materials. It contains information about the crystal structure, the degree of crystallinity of the sample, the temperature, the elastic strain as well as the defect density. The origin of the Raman scattering can be understand with the quantum mechanics. The scattering process between a photon and a phonon is described as an excitation of a phonon to a virtual state lower in energy than a real electronic state and the immediate de-excitation. After an extremely short period, in which the phonon is excited, it falls back either to the ground state or to an excited state (see Figure 1.1).

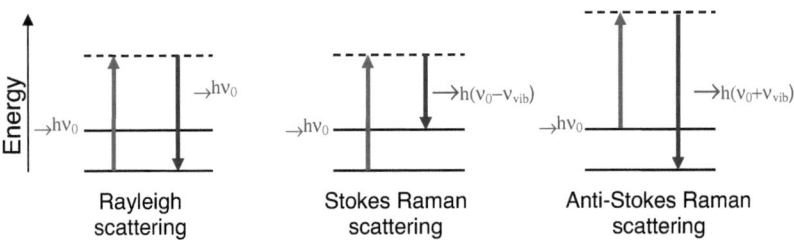

Figure 1.1 Vibrational energy level scheme for Rayleigh, Stokes andt Anti-Stokes scattering [Operating Manual for CRM 200 confocal Raman microscope by WITec].

Rayleigh scattering corresponds to light scattered at the frequency of the incident radiation. Raman scattering is shifted in frequency, and therefore energy, from the frequency of the irradiating light. Raman scattering can be divided into stokes scattering and anti-stokes scattering. In the case of stokes scattering the emitted light has less energy than the incident light while in the case of anti-stokes scattering the emitted photon has more energy than the incoming photon.

When a photon interacts with a phonon, the electrical Field $\vec{E}$ of the incident light induces a dipole moment $\vec{P}$ in the electronic distribution in molecules or atoms:

$$\vec{P} = \vec{\alpha} \cdot \vec{E}$$ (1.1)

The proportionality factor $\vec{\alpha}$ is the polarizability tensor which describes the response of the crystal to the electric field. The tensor is not a static value, but is dependent on the position of the atoms in the crystal. Calculating the Taylor expansion of α in x-direction while setting the equilibrium position x = 0 one can write:

$$\alpha(x) = \alpha(0) + \left(\frac{d\alpha}{dx}\right)_0 x + ...$$ (1.2)

Only linear terms are taken into account for small interatomic displacements. Looking at the phonon in the external electrical fields, one finds:

$$\vec{P} = \vec{\alpha} \cdot \vec{E} =$$ (1.3)

$$\alpha_0 E_0 \cos(\omega_0 t) + \frac{1}{2}\left[\left(\frac{\partial \alpha}{\partial x}\right)_0 x_= E_= \cos(\omega_= + \omega_m)t\right] + \frac{1}{2}\left[\left(\frac{\partial \alpha}{\partial x}\right)_0 x_= E_= \cos(\omega_= - \omega_m)t\right]$$

In equation (1.3) the first term represents the Rayleigh scattering, the second term the Stokes Stokes scattering and the third term the Anti-Stokes scattering. A crystal or a molecule is only Raman active if one of the components of the polarizability tensor changes during vibration.

Elastic strain due to mechanical stress affect the frequencies of the Raman-active phonons or molecular vibrations which lead to a shift of the Raman peak positions relative to the stress free peak position [38]. As a simple metaphor one can compare the process to the tuning of a guitar string. The tuning changes the elasitc strain and the mechanical stress in the guitar string which leads to a change of the frequency of the vibration and therefore would also lead to a shift of the positions of the Raman peaks. In the following section we will explain the underlying theory of measuring the elasitc strain and show how to convert the results into stress. Different methods to measure mechanical stresses in materials with a cubic crystal structure such as silicon will be presented. The stress resolution of the method depends on the stress sensitivity of the Raman peak and the signal-to-noise ratio of the measured signal.

Only certain types of lattice vibrations give rise to Raman scattering. The Raman tensors show which phonons are Raman active and what kind of orientation these phonons have. The Raman tensor can be determined from the crystal structure of the material. According to Louden [39], silicon has three Raman tensors. In the crystal coordinate system $x = [100]$, $y = [010]$ and $z = [001]$ they are:

$$\Delta_1 = \begin{bmatrix} 0 & 0 & 0 \\ 0 & 0 & d \\ 0 & d & 0 \end{bmatrix}, \Delta_2 = \begin{bmatrix} 0 & 0 & d \\ 0 & 0 & 0 \\ d & 0 & 0 \end{bmatrix}, \Delta_3 = \begin{bmatrix} 0 & d & 0 \\ d & 0 & 0 \\ 0 & 0 & 0 \end{bmatrix} \qquad (1.4)$$

The total scattering intensity of the phonon modes (I) for a given geometry can be determined from the Raman tensors, Δ_j, and the polarization vector of the incident (e_i) and scattered (e_s) light and is given by

$$I = C \sum_j \left| e_i \cdot \Delta_j \cdot e_s \right|^2 \qquad (1.5)$$

where C is a constant. Measuring in a back-scattering mode from a (001) surface the incident as well as the scattered light beam are normal to the surface. In this case Δ_1 and Δ_2 correspond to scattering from the transverse optical phonons (TO), polarized along the x- or the y-axis, respectively. Δ_3 correlates to scattering from the longitudinal optical phonons (LO), polarized along the z-axis. Whether the phonon is longitudinal or transversal depends on the surface from which scattering is observed. Measuring in a back-scattering mode from a (100) surface, Δ_1 corresponds to the LO phonon [16]. The phonons are triply degenerated which means that they have all the same wave number of approximately 520 cm^{-1}.

It follows from equation (1.5) that different modes can be measured depending on the orientation of the surface and the polarization direction of the light. Table 1.2 shows which of the modes can be obtained for a back-scattering experiment from a (001) surface, (110) surface, or a (11$\bar{2}$) surface.

Table 1.2 Polarization selection rules for back-scattering from a (001), (110) or a (11$\bar{2}$) surface.

Polarization		Visible		
e_i	e_s	Δ_1	Δ_2	Δ_3
Back-scattering from (001)				
(100)	(100)	-	-	-
(100)	(010)	-	-	x
(1$\bar{1}$0)	(1$\bar{1}$0)	-	-	x
(110)	(1$\bar{1}$0)	-	-	-
Back-scattering from (110)				
(1$\bar{1}$0)	(001)	x	x	-
(1$\bar{1}$0)	(1$\bar{1}$0)	-	-	x
(001)	(001)	-	-	-
Back-scattering from (11$\bar{2}$)				
(1$\bar{1}$0)	(111)	x	x	-
(1$\bar{1}$0)	(1$\bar{1}$0)	-	-	x
(111)	(111)	x	x	x

In a back-scattering configuration for a (001) surface only the longitudinal optical phonon mode is allowed regardless of the polarizations of the incident and the scattered light, whereas the back-scattering configuration of a (110) as well as a (11$\bar{2}$) surface allows the excitation of all three Raman modes [18]. In these latter two cases the longitudinal and the transversal phonons can be measured independently from each other. Measuring from a (111) surface all phonon modes are excited regardless of the polarization of the incident and scattered light [40].

1.4.1 Measuring in Conventional Backscattering Configuration

Strain caused by mechanical stress can influence the frequencies of the Raman mode [18, 38, 41, 42]. This section describes how a peak shift of silicon can be correlated to strains and stresses, respectively. In the absence of strain, the triply degenerated phonons of silicon lead to a peak at $\omega_0 = 520$ cm^{-1}. Strain changes the phonon vibrations and therefore can lift the degeneracy of the Raman frequencies due to a symmetry reduction of the Raman tensors [18, 43]. This symmetry reduction causes splitting and shifting of the Raman peak. Ganeasen et al. [38] showed that if strain is applied to a sample then the frequencies of the three optical phonon modes can be calculated by finding the eigenvalues, λ_j, j = 1,2,3, in the following equation

$$\begin{vmatrix} p\varepsilon_{11} + q(\varepsilon_{22} + \varepsilon_{33}) - \lambda_1 & 2r\varepsilon_{12} & 2r\varepsilon_{13} \\ 2r\varepsilon_{21} & p\varepsilon_{22} + q(\varepsilon_{11} + \varepsilon_{33}) - \lambda_2 & 2r\varepsilon_{23} \\ 2r\varepsilon_{31} & 2r\varepsilon_{32} & p\varepsilon_{33} + q(\varepsilon_{11} + \varepsilon_{22}) - \lambda_3 \end{vmatrix} = 0, \quad (1.6)$$

where $p = -1.85\,\omega_0^2$, $q = -2.31\,\omega_0^2$ and $r = -0.71\,\omega_0^2$ are the phonon deformation potentials (PDPs), which are material constants. ε_{ij} are the components of the strain tensor. The frequency of each mode in the presence of strain ω_j (j = 1, 2, 3) of the strained crystal is related to the unstrained frequency, ω_0, by the following relationship:

$$\omega_j^2 = \omega_0^2 - \lambda_j, \quad (1.7)$$

or for the approximation of small strains [18]:

$$\Delta\omega_j = \omega_j - \omega_0 \approx \frac{\lambda_j}{2\omega_0} \quad (1.8)$$

The polarization direction of each mode is given by the eigenvectors of equation (1.6).

The relation between the strain tensor ε in equation (1.6) and the stress tensor σ is given by Hooke's law

$$\begin{Bmatrix} \varepsilon_{11} \\ \varepsilon_{22} \\ \varepsilon_{33} \\ 2\varepsilon_{23} \\ 2\varepsilon_{13} \\ 2\varepsilon_{12} \end{Bmatrix} = \begin{bmatrix} s_{11} & s_{12} & s_{12} & 0 & 0 & 0 \\ s_{12} & s_{11} & s_{12} & 0 & 0 & 0 \\ s_{12} & s_{12} & s_{11} & 0 & 0 & 0 \\ 0 & 0 & 0 & s_{44} & 0 & 0 \\ 0 & 0 & 0 & 0 & s_{44} & 0 \\ 0 & 0 & 0 & 0 & 0 & s_{44} \end{bmatrix} \begin{Bmatrix} \sigma_{11} \\ \sigma_{22} \\ \sigma_{33} \\ \sigma_{23} \\ \sigma_{13} \\ \sigma_{12} \end{Bmatrix} \quad (1.9)$$

where $s_{11} = 7.68*10^{-2}$ Pa^{-1}, $s_{12} = -2.14*10^{-12}$ Pa^{-1} and $s_{44} = 12.7*10^{-12}$ Pa^{-1} are components of the elastic compliance tensor. This equation allows the calculation of the developing stresses from the measured strain. In the case of uniaxial stress along the [100] direction, equation (1.9) leads to the following nonzero components of the strain tensor: $\varepsilon_{11} = s_{11}\sigma_{11}$, $\varepsilon_{22} = s_{12}\sigma_{11}$, and $\varepsilon_{33} = s_{12}\sigma_{11}$. Solving (1.6) and (1.8) leads to the following results:

$$\Delta\omega_1 = \frac{\lambda_1}{2\omega_0} = \frac{1}{2\omega_0}(ps_{11} + 2qs_{12})\sigma$$

$$\Delta\omega_2 = \frac{\lambda_2}{2\omega_0} = \frac{1}{2\omega_0}[ps_{12} + q(s_{11} + s_{12})]\sigma \quad (1.10)$$

$$\Delta\omega_3 = \frac{\lambda_3}{2\omega_0} = \frac{1}{2\omega_0}[ps_{12} + q(s_{11} + s_{12})]\sigma$$

For small strains one can assume that the Raman tensors do not change. According to Table 1.2 when measuring in a backscattering mode from a (001) surface only the third Raman mode can be measured. Therefore, only $\Delta\omega_3$ has a nonzero value. Similar to the uniaxial case, calculations can be performed assuming biaxial stress ($\sigma_{xx} + \sigma_{yy}$). Then an increase of the Raman frequency corresponds to compressive stress while a decrease of the Raman frequency indicates tensile stress.

The majority of substrates for silicon based microelectronic devices have a (001) surface. As a consequence, the only measurable phonon is the longitudinal optical phonon and information is obtainable only from one parameter, namely for the out-of-plane component. As the stress tensor has six degrees of freedom, it is always necessary to make assumptions about the stress state. To show the importance of a realistic assumption for the stress tensor, three different stress states (hydrostatic, uniaxial along the x-axis, uniaxial along the z-axis) and their influence on the Raman signal are given [44]. In all three cases one supposes a Raman shift of 1 cm^{-1}. In the hydrostatic case, the corresponding values are:

$$\sigma = -540 \begin{pmatrix} 1/3 & 0 & 0 \\ 0 & 1/3 & 0 \\ 0 & 0 & 1/3 \end{pmatrix} MPa, \tag{1.11}$$

while if one assumes a uniaxial stress along the x-axis, the value changes to

$$\sigma = -450 \begin{pmatrix} 1 & 0 & 0 \\ 0 & 0 & 0 \\ 0 & 0 & 0 \end{pmatrix} MPa. \tag{1.12}$$

Equation (1.13) shows the case for uniaxial stress along the z-axis instead of the x-axis

$$\sigma = -900 \begin{pmatrix} 0 & 0 & 0 \\ 0 & 0 & 0 \\ 0 & 0 & 1 \end{pmatrix} MPa. \tag{1.13}$$

The large differences among the values clarify an important consequence. Either the performed measurement is made on samples where the shape of the tensor is already well known or one has to find a way to overcome the limitation of measuring only one phonon mode. In the following section we will explain how to measure stresses not only for the simple case where one phonon is excited but we will also show two different ways to overcome experimental limitations.

1.4.2 Off-axis Raman Spectroscopy

The complete stress tensor for an arbitrary crystal orientation can be determined by tilting the incident beam away from the normal axis while polarizing the incident as well as the scattered beam [45, 46]. This experimental set-up is called off-axis Raman spectroscopy. One has to keep in mind that in this set-up the laser is not focused and therefore the lateral resolution is low. So the method is suitable best for analyzing general stress states.

To detect all three Raman modes from a (001) surface of a silicon wafer, either the incident and/or the scattered light must have electric field components in all three x, y, and z directions to get signal from all three Raman tensors (see (1.5). As another requirement, incident and scattered light beams have to be polarized. If the incident light deviates from the surface normal, all three Raman modes are activated and can be measured. By adjusting the polarizer one is able to vary the proportion of each mode to the total intensity signal. The correct experimental set-up allows studying a selected phonon mode. Tuning the system allows one to determine all six components of the stress tensor.

In the off-axis Raman spectroscopy it is important to define several experimental parameters, which have to be under direct experimental control. In the here described case four different angles have to be known. The angle between the incident angle of the laser and the surface normal Φ is the first important parameter. This angle allows the detection of all three phonon mode due to the nonzero z-component. The orientation of the incident light (α) as well as the scattered light (β) adjusted by a polarizer must be known. Finally, the orientation of the sample (κ) with respect to the incident light is also of high importance (see Figure 1.2). Once the tilting angle (Φ) is fixed all the other angles are adjusted to optimize the sensitivity of the apparatus.

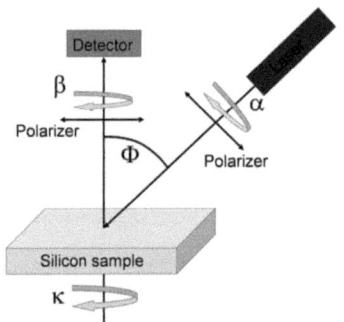

Figure 1.2 Simplified schematic of an experimental set-up for polarized off-axis Raman spectroscopy. α and β denote the orientation of the polarizer of the incident light and the scattered light, respectively. Φ is the tilting angle between the surface normal and the incident laser light. κ refers to the rotation angle of the silicon sample.

Due to the off-axis set-up of the laser, equation (1.5) has to be adjusted. The final signal intensity at the detector from a given phonon mode can be expressed by the following equation:

$$I = C(d_k^H B d_k). \qquad (1.14)$$

In this notation d_k is a pseudovector. Its components are the nonzero elements of the Raman tensor which only depend on the applied stress but not on the experimental set-up of the system. The superscript H denotes the complex-conjugate (Hermitian) transpose. B is the observation matrix which describes the directions of the vibrations. It depends only on the experimental set-up of the measuring system and on the material properties such as refractive index, but it does not depend on applied stress. C is a constant.

It is important to point out that in equation (1.14) the shape of the spectrum not only depends on the stress but is also a function of the experimental configuration, namely the observation directions given by the observation matrix B. B includes all the parameters which are under experimental control and must be calculated for every configuration (Φ, α, β, κ). By adjusting these parameters it is possible to selectively measure different phonon vibrations.

As an example the Raman spectra measured on a silicon wafer with a (001) surface orientation under biaxial stress is shown in Figure 1.3 [45]. The Raman measurements were taken parallel to the crystallographic axes. As one can see, it is possible to analyze different phonon vibrations in different observation directions. For visualizing the absolute shifts of the Raman peak position a spectrum from the unstressed wafer is plotted as well. The splitting between the two transversal optical phonons in the [100] and [010] direction and the longitudinal phonon in the [001] direction indicates biaxial stress in the wafer.

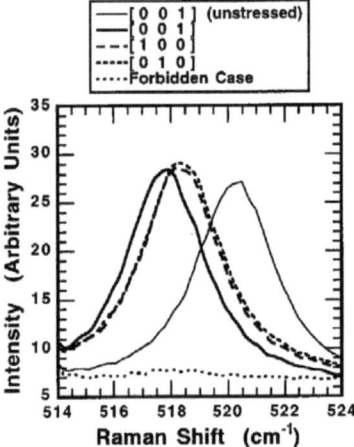

Figure 1.3 Raman spectra taken with observation directions along the crystallographic axes. All spectra show a significant shift to lower wavenumbers in comparison to the unstressed spectra which indicates tensile stress. Also a considerable splitting between the two in plane phonons ([100], [010]) and the out of plane phonon [001] can be seen [45].

In conclusion one can say that polarized off-axis Raman spectroscopy is a characterization method which allows the analysis of the complete stress tensor of a silicon wafer. The largest drawback of the method is the lateral resolution. This problem can be overcome by implementing the method for micro Raman spectroscopy.

1.4.3 Stress Tensor Analysis in Backscattering Raman Microscopy

We have seen that stress tensor analysis by means of Raman spectroscopy is generally possible, although the presented method does not provide local information about the stress state. To obtain local information about the stress is crucial for microelectronics as it can be the origin of hillocks, voids, and cracks [47]. Therefore a method is required which allows the measurement of stresses with a lateral resolution in the micrometer range of better. *Bonera et al.* presented a method to measure different components of the stress tensor with a spatial resolution of 1 μm by Raman spectroscopic technique [44, 48].

Equation (1.4) and (1.5) show, that one obtains only information from the longitudinal optical phonon from an (001) surface in an ideal back-scattering experiment. The two transversal optical phonons are not excited. In this case, the electric field of the laser light is perpendicular to the surface normal (see Figure 1.4 a)). If one uses an objective lens with a large numerical aperture (NA) all three phonons will be excited due to the fact that the incident laser light is no longer polarized perpendicular to the surface normal (see Figure 1.4 b)). The polarization vector also has a component in the z-axis. The higher the NA, then the higher the intensity from the transversal optical phonons will be.

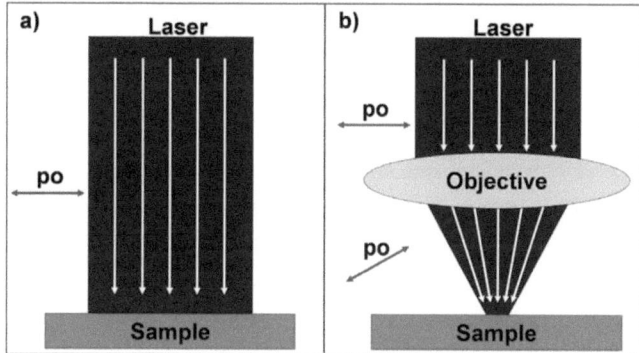

Figure 1.4 a) Ideal backscattering configuration. The polarization (po) of the incident laser light is perpendicular to the surface normal. Only the LO phonon is excited. b) Due to focusing the laser light the polarization vector (po) obtains a component parallel to the surface normal. Therefore all phonon modes are excited.

Polarizing the incident light as well as the scattered light offers the possibility to measure all three phonon modes separately. Using the Porto notation these, three cases can be written as:

$$z(y,y)\bar{z} \quad \rightarrow \omega_1$$
$$z(x,x)\bar{z} \quad \rightarrow \omega_2 \quad \quad (1.15)$$
$$z(y,x \otimes y)\bar{z} \rightarrow \omega_3$$

In the Porto notation, the first symbol, z, defines the direction of the in-coming light, while the last symbol, $\bar{z}$, refers to the direction of scattered light. The $\bar{z}$ implies that the scattered light is rotated 180° with reference to z. The first parameter within the parentheses describes the polarization of the incident light while the second symbol is the polarization direction of the scattered light. Equation (1.15) shows which experimental set-up is required for measuring a certain phonon mode. To study ω_3, which is the LO phonon, no polarizer is needed in the scattered beam. In this case, not only the LO phonon contributes to the total intensity but also one of the TO phonons. However, in comparison with the intensity of the LO phonon the portion of the TO phonons is negligible. Therefore, to get information from all three phonon vibrations one has to measure the same region with all three different polarization states of equation (1.15). In an unstressed area ω_1, ω_2 and ω_3 have the same frequency. In stressed areas the three frequencies are shifted. The total amount of the shift of every frequency depends on the actual stress state.

To calculate the stress state, one has to assume that the angles of the unit cells of the crystal do not change and the unit cell remains orthogonal. This assumption results in all nondiagonal values of the stress tensor to be set to zero

$$\sigma_{ij} = 0, \quad i \neq j. \quad \quad (1.16)$$

In most cases this assumption is a simplification of the true stress state. Nevertheless, it allows one to measure three of the six components of the stress tensor instead of one out of six. It is convenient to split the stress tensor into a hydrostatic (σ_h) and a traceless (σ_t) component.

$$\sigma = \sigma_h + \sigma_t \qquad (1.17)$$

Using equation (1.17) simplifies the secular matrix for the strained crystal (see equation (1.6)) to the following three linear equations:

$$\Delta\omega_j = \frac{(p+2q)(s_{11}+2s_{12})}{2\omega_0}\alpha'\cdot\sigma_h\cdot\alpha + \frac{(p-q)(s_{11}-s_{12})}{2\omega_0}\alpha'\cdot\sigma_t\cdot\alpha \qquad (1.18)$$

where $\alpha = x$, y, and z; s_{11} and s_{12} are elements of the compliance tensor (see equation (1.9)); p and q are phonon deformation potentials. Equation (1.18) allows determining the stress tensor with a lateral resolution depending on the wave length of the laser and the numerical aperture of the lens.

In conclusion, the above section presents a method to measure stress induced Raman shifts from a silicon surface. The spatial resolution is in the submicrometer range. The method allows to measure different phonon polarizations which belong to different directions in space.

1.5 References

1. G.E. Moore, *Cramming more components onto integrated circuits (Reprinted from Electronics, pg 114-117, April 19, 1965)*. Proceedings of the Ieee, 1998. **86**(1): p. 82-85.
2. J.R. Lloyd, *Electromigration and mechanical stress*. Microelectronic Engineering, 1999. **49**(1-2): p. 51-64.
3. M.A. Korhonen, P. Borgesen, K.N. Tu, and C.Y. Li, *Stress evolution due to electromigration in confined metal lines*. Journal of Applied Physics, 1993. **73**(8): p. 3790-3799.
4. Y. Toivola, J. Thurn, R.F. Cook, G. Cibuzar, and K. Roberts, *Influence of deposition conditions on mechanical properties of low-pressure chemical vapor deposited low-stress silicon nitride films*. Journal of Applied Physics, 2003. **94**(10): p. 6915-6922.
5. J.H. Park, J.H. An, Y.J. Kim, Y.H. Huh, and H.J. Lee, *Tensile and high cycle fatigue test of copper thin film*. Materialwissenschaft und Werkstofftechnik, 2008. **39**(2): p. 187-192.
6. D.H. Alsem, O.N. Pierron, E.A. Stach, C.L. Muhlstein, and R.O. Ritchie, *Mechanisms for fatigue of micron-scale silicon structural films*. Advanced Engineering Materials, 2007. **9**(1-2): p. 15-30.
7. R. Ghosh, D. Basak, and S. Fujihara, *Effect of substrate-induced strain on the structural, electrical, and optical properties of polycrystalline ZnO thin films*. Journal of Applied Physics, 2004. **96**(5): p. 2689-2692.
8. R. Spolenak, W.L. Brown, N. Tamura, A.A. MacDowell, R.S. Celestre, H.A. Padmore, B. Valek, J.C. Bravman, T. Marieb, H. Fujimoto, B.W. Batterman, and J.R. Patel, *Local plasticity of Al thin films as revealed by X-ray microdiffraction*. Physical Review Letters, 2003. **90**(9).
9. N. Tamura, A.A. MacDowell, R. Spolenak, B.C. Valek, J.C. Bravman, W.L. Brown, R.S. Celestre, H.A. Padmore, B.W. Batterman, and J.R. Patel, *Scanning*

X-ray microdiffraction with submicrometer white beam for strain/stress and orientation mapping in thin films. Journal of Synchrotron Radiation, 2003. **10**: p. 137-143.

10. M.A. Phillips, R. Spolenak, N. Tamura, W.L. Brown, A.A. MacDowell, R.S. Celestre, H.A. Padmore, B.W. Batterman, E. Arzt, and J.R. Patel, *X-ray microdiffraction: local stress distributions in polycrystalline and epitaxial thin films*. Microelectronic Engineering, 2004. **75**(1): p. 117-126.

11. J. Nucci, S. Kramer, E. Arzt, and C.A. Volkert, *Local strains measured in Al lines during thermal cycling and electromigration using convergent-beam electron diffraction*. Journal of Materials Research, 2005. **20**: p. 1851-1859.

12. J. Bauch, J. Brechbuhl, H.J. Ullrich, G. Meinl, H. Lin, and W. Kebede, *Innovative analysis of X-ray microdiffraction images on selected applications of the Kossel technique*. Crystal Research and Technology, 1999. **34**(1): p. 71-88.

13. R.R. Keller, A. Roshko, R.H. Geiss, K.A. Bertness, and T.P. Quinn, *EBSD measurement of strains in GaAs due to oxidation of buried AlGaAs layers*. Microelectronic Engineering, 2004. **75**(1): p. 96-102.

14. M. Kamaya, A.J. Wilkinson, and J.M. Titchmarsh, *Measurement of plastic strain of polycrystalline material by electron backscatter diffraction*. Nuclear Engineering and Design, 2005. **235**(6): p. 713-725.

15. Q. Ma, S. Chiras, D.R. Clarke, and Z. Suo, *High-Resolution Determination Of The Stress In Individual Interconnect Lines And The Variation Due To Electromigration*. Journal of Applied Physics, 1995. **78**(3): p. 1614-1622.

16. I. DeWolf, *Micro-Raman spectroscopy to study local mechanical stress in silicon integrated circuits*. Semiconductor Science and Technology 1995. **11**: p. 139-154.

17. F. Cerdeira, Buchenau.Cj, M. Cardona, and F.H. Pollak, *Stress-induced shifts of first-order Raman frequencies of diamond and zinc-blende-typ semiconductors*. Physical Review B, 1972. **5**(2): p. 580-&.

18. E. Anastassakis, A. Pinczuk, E. Burstein, F.H. Pollak, and M. Cardona, *Effect of Static Uniaxial-Stress on The Raman-Spectrum of Silicon (Reprinted From Solid-State Commun, Vol 8, Pg 133-138, 1970)*. Solid State Communications, 1993. **88**(11-12): p. 1053-1058.

19. G. Abstreiter, *Micro-Raman spectroscopy for characterization of semiconductor-devices*. Applied Surface Science, 1991. **50**(1-4): p. 73-78.
20. A.K.S. V.T. Srikar, M. Selim Ünlü, Bennett B. Goldberg, Mark Spearing, *Micro-Raman Measurement of Bending Stresses in Micromachined Silicon Flexures*. Journal of Microelectromechanical Systems, 2003. **Vol. 12**(No.6): p. 779-787.
21. M. Moskovits, *Surface-enhanced spectroscopy*. Reviews of Modern Physics, 1985. **57**(3): p. 783-826.
22. S.M. Nie and S.R. Emery, *Probing single molecules and single nanoparticles by surface-enhanced Raman scattering*. Science, 1997. **275**(5303): p. 1102-1106.
23. L. Zhu, C. Georgi, M. Hecker, J. Rinderknecht, A. Mai, Y. Ritz, and E. Zschech, *Nano-Raman spectroscopy with metallized atomic force microscopy tips on strained silicon structures*. Journal of Applied Physics, 2007. **101**(10).
24. C.V. Raman, *A change of wavelength in light scattering (Reprinted from Nature, vol 121, pg 619, 1928)*. Current Science, 1998. **74**(4): p. 381-382.
25. G. Landsberg and L. Mandelstam, *A new occurrence in the light diffusion of crystals*. Naturwissenschaften, 1928. **16**: p. 557-558.
26. B. Trumpy, *Rotational Raman spectrum of O/sub 2/ at high pressures*. Zeitschrift fur Physik, 1933: p. 282-288.
27. S.C. Sirkar, *Rotational raman scattering in benzene vapour*. Indian Journal of Physics, 1935: p. 295-298.
28. M. Linhaid, *Liquid ammoniac as a solvent for inorganic compounds - III Vapour pressure measurement*. Zeitschrift fur Physikalische Chemie-Abteilung a-Chemische Thermodynamik Kinetik Elektrochemie Eigenschaftslehre, 1936. **175**(6): p. 438-458.
29. S.S. Mitra, O. Brafman, W.B. Daniels, and R.K. Crawford, *Pressure-Induced Phonon Frequency Shifts Measured By Raman Scattering*. Physical Review, 1969. **186**(3): p. 942-&.
30. Anastass.E, A. Pinczuk, E. Burstein, F.H. Pollak, and M. Cardona, *Effect of static uniaxial stress on Raman spectrum of silicon*. Solid State Communications, 1970. **8**(2): p. 133-&.

31. M. Delhaye and P. Dhamelincourt, *Raman microprobe and microscope wich laser excitation*. Journal of Raman Spectroscopy, 1975. **3**(1): p. 33-43.
32. S.A. Lyon, R.J. Nemanich, N.M. Johnson, and D.K. Biegelsen, *Microstrain in laser-crystallized silicon islands on fused-silica*. Applied Physics Letters, 1982. **40**(4): p. 316-318.
33. S.R.J. Brueck, B.Y. Tsaur, J.C.C. Fan, D.V. Murphy, T.F. Deutsch, and D.J. Silversmith, *Raman measurements of stress in silicon-on-sapphire device structures*. Applied Physics Letters, 1982. **40**(10): p. 895-898.
34. A. Draaijer, M.H. Vandijk, and P.M. Houpt, *Confocal laser scanning microscopy in combination with Raman-spectrometry for 3D-Raman mapping*. Transactions of the Royal Microscopical Society : New Series, Vol 1, 1990. **1**: p. 455-458.
35. G.J. Puppels, F.F.M. Demul, C. Otto, J. Greve, M. Robertnicoud, D.J. Arndtjovin, and T.M. Jovin, *Studying single living cells and chromosomes by confocal Raman Microspectroscopy*. Nature, 1990. **347**(6290): p. 301-303.
36. M. Moskovits, *Surface-roughness and enhanced intensity of Raman-scattering by molecules adsorbed on metals*. Journal of Chemical Physics, 1978. **69**(9): p. 4159-4161.
37. J. Steidtner and B. Pettinger, *Tip-enhanced Raman spectroscopy and microscopy on single dye molecules with 15 nm resolution*. Physical Review Letters, 2008. **100**(23).
38. S. Ganesan, Maradudi.Aa, and J. Oitmaa, *A Lattice Theory Of Morphic Effects In Crystals Of Diamond Structure*. Annals of Physics, 1970. **56**(2): p. 556-&.
39. R. Loudon, *Raman Effect in Crystals*. Advances in Physics, 1964. **13**(52): p. 423-&.
40. S. Narayanan, S.R. Kalidindi, and L.S. Schadler, *Determination of unknown stress states in silicon wafers using microlaser Raman spectroscopy*. Journal of Applied Physics, 1997. **82**(5): p. 2595-2602.
41. E. Anastassaskis, E. Burstein, *Morphic Effects .2. Effects of External Forces On Frequencies Of Q=O Optical Phonons*. Journal of Physics and Chemistry Of Solids, 1971. **32**(3): p. 563-&.

42. E. Anastassaskis, E. Burstein, *Morphic Effects .1. Effects of External Forces On Photon-Optical Phonon Interactions.* Journal of Physics and Chemistry Of Solids, 1971. **32**(2): p. 313-&.
43. I. DeWolf, J. Vanhellemont, A. Romanorodriguez, H. Norstrom, and H.E. Maes, *Micro-Raman Study Of Stress-Distribution In Local Isolation Structures And Correlation With Transmission Electron-Microscopy.* Journal of Applied Physics, 1992. **71**(2): p. 898-906.
44. E. Bonera, M. Fanciulli, and D.N. Batchelder, *Combining high resolution and tensorial analysis in Raman stress measurements of silicon.* Journal of Applied Physics, 2003. **94**(4): p. 2729-2740.
45. G.H. Loechelt, N.G. Cave, and J. Menendez, *Polarized off-axis Raman spectroscopy: A technique for measuring stress tensors in semiconductors.* Journal of Applied Physics, 1999. **86**(11): p. 6164-6180.
46. G.H. Loechelt, N.G. Cave, and J. Menendez, *Measuring The Tensor Nature Of Stress In Silicon Using Polarized Off-Axis Raman-Spectroscopy.* Applied Physics Letters, 1995. **66**(26): p. 3639-3641.
47. S.M. Hu, *Stress-related problems in silicon technology.* Journal of Applied Physics, 1991. **70**(6): p. R53-R80.
48. E. Bonera, M. Fanciulli, and D.N. Batchelder, *Raman spectroscopy for a micrometric and tensorial analysis of stress in silicon.* Applied Physics Letters, 2002. **81**(18): p. 3377-3379.

2 *In Situ* Compression Tests on Micron Sized Silicon Pillars by Raman Microscopy-Stress Measurements and Deformation Analysis

This chapter presents results of *in situ* compression tests performed on micro-sized silicon pillars. This novel method can be performed at high data acquisition rates and has the advantage to give information about local compressive stresses as well as microstructural changes occurring during the compression test. As the applied load causes well defined uniaxial stresses, it is an ideal system to compare stresses calculated with a load cell and stresses calculated with the data deriving from Raman microscopy. This gives a good idea about the accuracy of Raman microscopy as a tool for measuring strains or stresses. This work has been published as a full length article:

Wasmer, K., Wermelinger, T., Bidiville A., Spolenak, R. and Michler, J.; *In Situ Compression Tests on Micron Sized Silicon Pillars by Raman Microscopy-Stress Measurements and Deformation Analysis*, Journal of Materials Research 23 (2008) 3040-3047

2.1 Abstract

Mechanical properties of silicon are of high interest to the micro-electro-mechanical systems community as it is the most frequently used structural material. Compression tests on 8 µm diameter silicon pillars were performed under a micro-Raman setup. The uniaxial stress in the micro-pillars was derived from a load cell mounted on a micro-indenter and from the Raman peak shift. Stress measurements from the load cell and from the micro-Raman spectrum are in excellent agreement. The average compressive failure strength measured in the middle of the micro-pillars is 5.1 GPa. Transmission electron microscopy investigation of compressed micro-pillars showed cracks at the pillar surface or in the core. A correlation between crack formation and dislocation activity was observed. The authors strongly believe that the combination of nanoindentation and micro-Raman spectroscopy allowed detecting cracks prior to failure of the micro-pillar, which also allowed an estimation of the in-plane stress in the vicinity of the crack tip.

2.2 Introduction

Characterization of the material properties of silicon (Si) in the micro- and sub-micrometer range has become of particular interest in the last couple of years. This is due to the fact that this material is widely employed in thin film devices and Micro-Electro-Mechanical Systems (MEMS), where size effects have been observed and examined by many researchers. This is especially the case for Si in the sub-millimeter [1-3] and micrometer [4] ranges. In these studies, mechanical properties were obtained from either bending or tensile tests. Recently, micro-compression tests on Si micro-pillars with diameters ranging from 1 to 50 μm were employed to characterize the material properties and to study the size effects [5]. The available literature on micro-pillars testing focuses on brittle rupture of Si rather than the general stress-strain (σ-ε) behavior. On the other hand, micro-Raman spectroscopy (μRS) is often used to study and/or measure the local mechanical stresses in silicon wafers and silicon integrated circuits [6]. The stress resolution of Raman microscopy in the case of silicon is ~25 MPa [7]. Furthermore, it is applied to investigate the microstructure of silicon. Indentation or scratching operations lead to phase transformations. The type of occurring phases can be determined with Raman microscopy due to their unequal Raman spectra [8-12]. Another microstructural aspect which is influencing the Raman spectra is the grain size. Iqbal and Veprek showed that below a certain threshold (~10 nm) the grain size of poly crystalline silicon can be determined from the according Raman spectra [13]. This work presents for the first time the combination of a µRS with a homemade micro-indenter in order to characterize the stress state localized in silicon micro-pillars during testing and to investigate the main deformation mechanisms *in situ*. The combination of these two techniques allows correlation of the mechanical properties with microstructural changes within the focus of the Raman laser during deformation.

2.3 Experiments and Materials

Samples of 10x10 mm^2 were cut from a 4 inch, 525 µm thick (001)-oriented Si wafer. Cylindrical single crystalline silicon pillars were fabricated by standard photolithography techniques on the sample. A Heidelberg DWL200 (Heidelberg Instrument GmbH, Heidelberg, Germany) direct laser writer was used to transfer the computer pre-defined design (with a square array of 10 µm diameter circles) on the photo-resist (Shipley Microposit S1800 series, Shipley Europe Ltd., Conventry, UK) coated wafer. The samples were obtained from silicon wafers just coated by a photoresist layer. After irradiation, and resist development, the wafers were anisotropically etched. To do this operation, a pulsed room temperature process (so-called Bosch process [14]) was applied. This consists of a cyclic process consisting of a few seconds of Si etching (by SF_6) followed by a few seconds sidewall protection (by C_4F_8), in the same plasma Alcatel 601 etcher (Alcatel Vacuum Technology, Annecy, France). When the etching process was completed, the residual photoresist was removed by a specific wet remover (Microposit Remover 1165, Shipley Europe Ltd., Conventry, UK), and further cleaned by an oxygen plasma in a Branson IPC 2000 Plasma System. The method is fully described elsewhere [5].

A typical pillar with its dimension is shown in Fig. 1. From this figure it is seen that, the diameter of the pillar varies slightly along its length due to the photolithography technique. Consequently, the geometry of all investigated pillars was measured by

Scanning Electron Microscopy (SEM) prior to deformation. The diameters were measured, at about half of the height of the micro-pillars, at the approximate location of the laser beam of the Raman microscope (Ø 7.3 µm from Figure 2.1). The length of all pillars was very uniform.

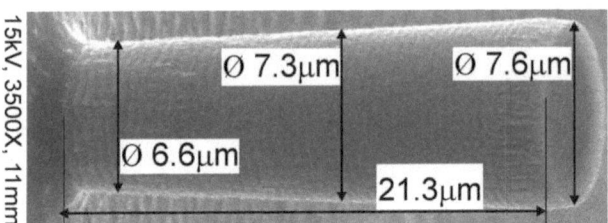

Figure 2.1 SEM picture of a typical micro-pillar used for the micro-compression test. The arrow indicates the approximate position of the diameter measurement.

Their geometry and mean strength measurements (taken from [5]) are summarized in Table 2.1.

Table 2.1 Geometry of the micro-pillars and their mean strength measurement with its respective ± 2 standard deviation (95% confidence interval)

Location of diameter measurement	Sample diameter [µm]	Length [µm]	Aspect ratio (length/diameter)	Mean compressive fracture strength [GPa]
Fracture	6.6 ± 0.06	21.3	3.3	-7.0 ± 0.3 from [5]
Middle	7.3 ± 0.06	21.3	2.9	-5.1 ± 0.4

The surfaces of the micro-pillars are not perfectly smooth as displayed in Figure 2.1, which is inherent to the deep reactive ion etching process employed [5]. The Raman microscope employed was a Confocal Raman Microscope CRM 200 (WITec GmbH, Ulm) with a helium cadmium laser with a wavelength of 442 nm. All measurements were performed in backscattering mode without any filter. A 20x objective with a numerical aperture of 0.4 was used.

The micro-compression experiments were performed using a homemade instrumented micro-indentation device developed to operate inside a SEM. This apparatus, fully described elsewhere [15], is based on a load cell (maximum load P_{Max} of 500 mN) fixed on a piezo-actuated positioning stage. In contrast to the description given in [15], a second stick slip stage was added to allow full Cartesian positioning of the specimen with respect to the indenter tip. To identify the micro-pillars, the optical microscope of the Raman spectrometer was used. The load axis was inclined by 84° to the microscope axis. The entire setup is shown in Figure 2.2.

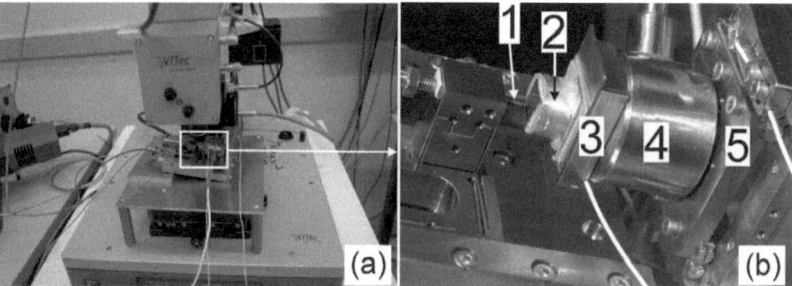

Figure 2.2 Photographs showing (a) the experiments under the micro-Raman and (b) the set-up of the compression experiment with (1) diamond tip, (2) specimen holder, (3) y-positioning stage, (4) load cell and (5) x-positioning stage.

The compression load was applied in the [001] direction whereas the laser spotted the pillar at 90° along the [010] direction. In such set-up, the only the in-plane phonon give rise to the Raman signal. Therefore, only the in-plane component of the stress tensor is measured. The corresponding shift of the Si Raman peak exhibits a stress sensitivity of 500 MPa/cm^{-1} [6, 16].

Compression tests were carried out using a diamond flat punch with a diameter of approximately 10 μm. The experiments were performed under constant displacement rate. The load on the sample, the displacement of the tip via the stack piezo and the micro-Raman spectra were recorded simultaneously. The engineering stress, σ, was defined according to $\sigma = P/A$, where P is the applied load and A the initial cross-section determined from SEM images. The cross-sectional area A was calculated from the diameter in the middle of the height of the sample, as illustrated in Figure 2.1, to correspond to the micro-Raman measurements.

To investigate the microstructure and the deformation mechanisms of the micro-pillars (phase transformations and/or crack initiation and propagation arising during the micro-mechanical tests), TEM lamellae of pillars after loading up to 200 mN were prepared by Focused Ion Beam (FIB, FEI Strata 235, FEI Europe B.V., Zürich, Switzerland) and observed with a Philips CM30 Philips LaB6 TEM (Philips Europe, Eindhoven, Netherland) using 300 kV acceleration voltage. In order to distinguish the deformation process at the center of the pillar and on its side, one lamella was prepared in each location.

Forty micro-compression tests of two types were conducted. Firstly, twenty tests were loaded in purely elastic regime and unloaded, without holding time. The first ten were loaded up to 140 mN ($\sigma = -3.3$ GPa) whereas the maximum load for the next ten experiments was 200 mN ($\sigma = -4.6$ GPa). Secondly, twenty tests were loaded until rupture. For each experiment the loading and unloading rate was 1.0 mN/s and the force via the load cell and Raman measurements were recorded every 0.2 sec.

2.4 Results

The results of the twenty tests until failure are summarized in Table 1, where the mean compressive fracture strength in the middle of the pillar is -5.1 GPa with a standard deviation of ± 0.4. For the same tests set, the Weibull parameters were calculated and the Weibull modulus m equals 15.6 and the characteristic strength σ_0 equals 5.3 GPa.

Figure 2.3 (a) displays a pillar that was compressed up to -4.6 GPa and unloaded. Inspection of this figure indicates that no crack or apparent damage is observed. The Raman spot was added to compare the spot with the pillar sizes. The corresponding plot of the Raman shift during the loading-unloading experiment is presented in Figure 2.3 (b). The Raman spectrum is given only between 500 and 550 cm^{-1} since no other peak is of further interest outside this range. The stresses, given in GPa, were derived from the load cell. The shift in the Raman spectra due to the compressive stress is evident. No difference is observed between Raman spectra obtained, at identical stresses, during the loading and unloading sequence. This is also corroborated by the stress-displacement curve in Figure 2.3 (c). This figure compares the compressive stress derived from (a) the load cell, based on the engineering stress, and (b) the Raman shift.

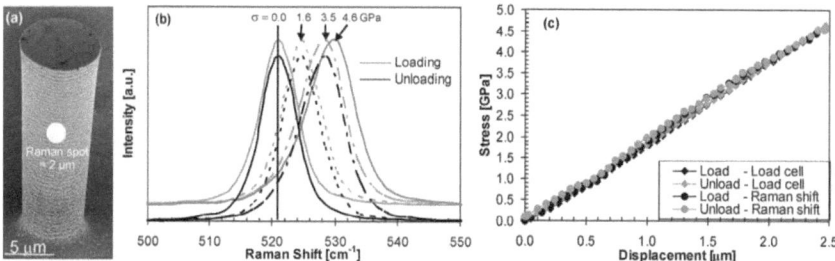

Figure 2.3 Micro-indenter and micro-Raman results of an elastic micro-pillar compression test. (a) Micro-pillars subjected to a compressive stress of -4.6 GPa with the Raman spot. (b) Raman shift measured during loading – unloading with a velocity of 0.7 mN/s and a maximum load of 200 mN. The spectra intensity has been adjusted to facilitate the visual comprehension. (c) Comparison of the stress-displacement curve between both methods.

The stresses derived from both methods are in very good agreement since the discrepancy between both methods throughout the test is less than 3%. Finally, the lack of visible differences between the loading and unloading curves demonstrates that pillars compressed up to a load of -4.6 GPa behave purely elastically.

In three cases of the compression tests, a shoulder on the low energy side of the silicon Raman peak was observed as illustrated in Figure 2.4 by two examples. In Figure 2.4(a), the shoulder appears in the spectrum at a compressive stress of -1.87 GPa and disappears during the measurement when -2.47 GPa were applied. The peak position of the shoulder moves to lower wavenumbers, as indicated by the arrow, with increasing stress. The shape of the shoulder at a stress of -2.45 GPa and the shape of the curve at a stress of -2.47 GPa look very different although the applied stresses are almost equal. One has to keep in mind that 2 seconds have passed between the measurements of the two curves. In contrast, the shoulder in Figure 2.4 (b) starts to be visible at -1.17 GPa and vanishes at -2.10 GPa. The arrow highlights that the peak position of the shoulder increases with increasing compressive stress. In both tests, the shoulder appears and disappears during the loading part of the test, although it moves in an opposite direction between Figure 2.4 (a) and 4(b).

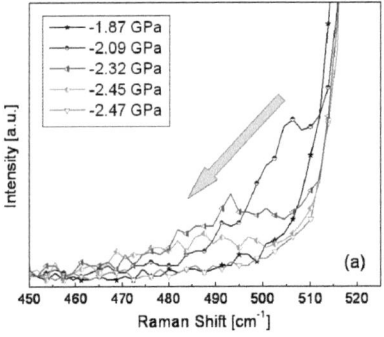 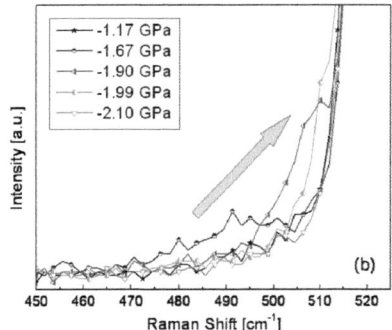

Figure 2.4 Occurrence of a shoulder on the low energy side of the Raman peak in two different tests. (a) Shoulder appears between -1.87 GPa and -2.47 GPa stress. The peak position of the shoulder moves to slower wavenumbers. (b) Shoulder appears between -1.17 GPa and -2.10 GPa stress. The peak position of the shoulder moves to higher wavenumbers.

Figure 2.5 and Figure 2.6 show two TEM lamellae machined with the [110] crystallographic direction normal to the foil. The lamella in Figure 2.5 (a) was cut in the middle of the pillar (see Figure 2.5 (b)) and shows several interesting features. The most obvious is a large crack initiated at approximately 600 nm under the surface and propagating to the center of the pillar to a maximal depth of 4.4 μm in the pillar. This crack was already visible from the SEM image during the FIB preparation. It can be seen from the inset (c) in Figure 2.5 that additional small cracks with different orientation exist along an array of few dislocations. As the pillars were made on a (001) wafer, their axis belongs to a [001] direction and so the plane of the lamella is {110}. From these geometrical considerations, it appears that the small cracks are along {111} planes, as expected from free surface energy. As no dislocation was observed on other parts of this lamella, it is supposed that that their nucleation was triggered by crack interaction. The diffraction pattern (Figure 2.5 (d)) demonstrates that no permanent phase transformation can be seen.

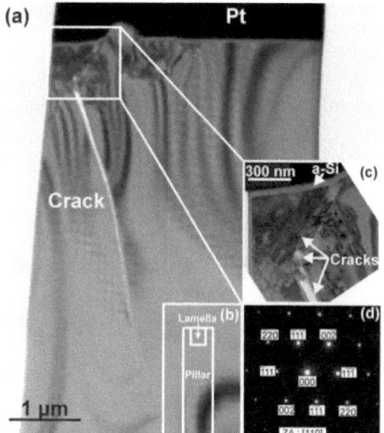

Figure 2.5 TEM bright field ([110] projection) images of a FIB lamella made in the middle of a pillar subjected to a compress stress of -4.6 GPa. (a) Overview of the pillar with a crack starting approximately 800 nm under the surface and with a platinum layer at the surface. (b) Location of the FIB lamella in the middle of the micro-pillar. (c) The inset illustrates the location of the long cracks as well as small cracks along the few dislocations visible. (d) The diffraction pattern taken from the inset shows that no phase transformation occurred in this region during the experiments.

The lamella cut at the edge of the pillar presented in Figure 2.6 contains only a large crack initiated at the surface with a depth of 800 nm but is completely free from dislocations. Similarly to Figure 2.5, the crack was observed previously from the SEM image taken during the FIB preparation. Figure 2.6 (a) shows the edge of the pillar prepared according to Figure 2.6 (b) and the roughness coming from the preparation by the photolithography technique. No dislocation is seen, even at stress-concentrating features like the edges of the crack, corners or roughness on the pillar wall and evidence of this is shown in Figure 2.6 (d). A small (< 50 nm) amorphous layer on top of the pillar coming from the protective platinum layer deposition [17, 18] is observed in both figures. In Figure 2.6 (c), it is observed that the platinum layer is visible on the side of the crack but has been milled away at the bottom of the crack. This indicates that the crack was created during the experiment and that it was partly enlarged by the ion beam during the TEM sample preparation.

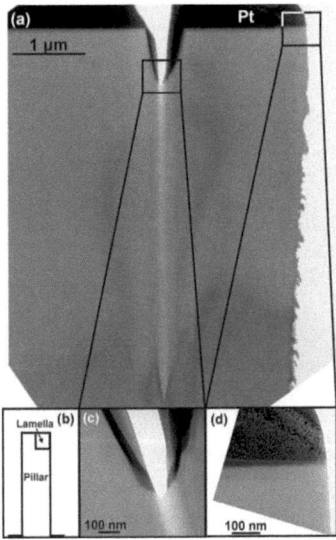

Figure 2.6 TEM bright field ([110] projection) of the lamella made at the edge of a pillar subjected to a compression stress of -4.6 GPa. (a) Overview of a dislocation-free pillar milled on the side of the pillar with a crack starting at the surface. It shows as well the platinum layer at the surface and the roughness of the side of the pillar. (b) Location of the FIB lamella at the edge of the micro-pillar. (c) The inset illustrates the enlargement of the crack presented and that no dislocation is seen. (d) The inset shows the edge of the pillar where no dislocation is visible.

2.5 Discussion

In situ SEM observations of micro-compression tests revealed that the micro-pillars used in this work break mainly by brittle fracture without buckling. Only some pillars exhibited visible crack formation before rupture [5]. Furthermore, earlier observations showed that pillars partly sank into the substrate during testing which is in agreement with an elastic model of the compression behavior [19].

Compared to the results given by Moser *et al.* [5], the engineering stress at fracture in the current experiment given in Table 1 is lower by 30%. This discrepancy is solely due to the difference in diameter employed to calculate the engineering stress. Moser *et al.* [5] used the pillar diameter at the location of its fracture, which is at its base (Ø 6.6 µm). In this work, the diameter at the position of the laser spot was taken for stress calculation that is Ø 7.3 µm based on Figure 2.1 and Figure 2.3 (a). Strain measurements derived from the displacement of the stack piezo are influenced by the compliance of the load cell as well as the sink-in of the pillar into the substrate [19] which has already been addressed by Moser *et al.* [5]. These two disadvantages make the derivation of the strain in the pillars from the load-displacement data difficult. Thus, effects that depend on an accurate strain measurement will not be discussed. Moreover, one should bear in mind that different volumes are probed by load cell measurements versus Raman microscopy. With the load cell, the entire volume of the pillar contributes to the signal. In contrast, the Raman microscopy measures a relatively small roughly cylindrical volume close to the surface on the side of the pillar (Ø 2 µm) and a depth of 250 nm.

The stresses calculated from the load cell data and from the Raman method in Figure 2.3 (b) and Figure 2.3 (c) are in very good agreement as they differ by less than 3%. This divergence is negligible considering (a) the uncertainties in the Raman measurement (the precision of the peaks is in the order of 0.05 cm^{-1} wave numbers), (b) the error in the load cell measurements of the micro-indenter, which is around 1 mN, (c) the error in phonon shifts from the thermal expansion due to sample heating (≈ 0.025 cm^{-1}/K [16]), (d) the error of defocusing, which is less than ± 0.05 cm^{-1} in the range of -1 μm < z < 1 μm [16] and finally (e) the accuracy in positioning the laser exactly where the diameter was measured for the stress calculation. Concerning the latter and assuming a positioning error of 1 μm along the height of the pillar in Figure 2.1, the cross-section varies by 2.5 %. Figure 2.3 (b) clearly demonstrates that the stress relaxation is reversible. This is supported by Figure 2.3 (c) and the TEM investigations in Figure 2.5 and Figure 2.6 which demonstrate that no phase transformation but only few cracks and dislocations exist within the compressed pillars to stresses up to -4.6 GPa. Hence, it can be concluded that its behavior is mainly elastic with very little plastic deformation.

The shoulders on the low energy side of the Raman silicon peak shown in Figure 2.4 are visible in only three experiments. Two examples of such behavior are displayed in Figure 2.4. It is evident from this figure that the behavior of the shoulders differs in terms of point of appearance and duration but the band lays within 490 – 520 cm^{-1}. Possible origins of these bands, taking into account the shift due to the compressive stress, could be assigned to either the formation of Si nanocrystals [20, 21], Si-IV (hexagonal diamond structure) phases [8]-13, 24] or regions with extensive cracking [8-11, 22]. These mechanisms have been observed to result in similar peaks as shown in Figure 2.4.

As the formation of nanocrystalline Si requires a large amount of plastic deformation and all specimens failed in a brittle manner, this explanation can be discarded. Moreover, phase transformation from Si-I to Si-IV can occur directly from Si-I through twin intersections and depends on the density of twins in the shear deformed material [11, 23], this explanation is as well neglected based on the TEM analyses in Figure 2.5 and Figure 2.6. Consequently, only regions with extensive cracking seem a possible explanation for the shoulder observed.

Analyzing in detail the various results presented, all pillars deformed purely elastically up to a stress of -1.17 GPa (prior to the appearance of a shoulder in the Raman spectrum). Real-time *in situ* SEM test observations and the load-displacement curves showed that no irreversible deformation or cracking occurs for pillars unloaded just before failure [5]. However, careful post-examinations showed that in some pillars longitudinal cracking exists. As shown from the TEM investigation, such cracks are likely to occur at the intersection of slip bands or nucleate at surface irregularities which is consistent with the findings of Lloyd et al. [24]. The crack tip is surrounded by a small zone of extremely high tensile stress, which can be determined from the stress intensity factor K, according to $K = \sigma \cdot \sqrt{\pi \cdot a}$, where σ and a are the engineering stress and the crack length, respectively. The theoretical tensile strength (σ_{th}) of silicon ranges between 13 and 18.5 GPa depending on the crystallographic direction, assuming a rule of thumb $\sigma_{th} = E/10$ and making use of $E_{001} = 130$ GPa and $E_{111} = 185$ GPa [25]. This is consistent with the literature where Hoffmann et al. [26] have measured a bending failure strength of 12 GPa for nanowires grown on a [111] silicon substrate by the vapor–liquid–solid process with diameters ranging from 100 to 200 nm and a length of 2 μm. Namazu et al. [4] have also found an average bending strength of 17.5 GPa for micromachined silicon beams along [111] direction of 6 μm length and around 250 nm in diameter. Moreover, molecular dynamics simulations indicate even higher strengths of up to 40 GPa are possible [27]. If such a crack propagates through the laser focus spot, the Raman signal is undoubtedly affected. Actually, tensile stresses will shift the Raman peak to lower wavenumbers which is in agreement with Fig. 4. By combining the Raman with the TEM observations, one could speculate that the shoulder originates from a crack growing through the laser spot. Stress evaluation from the difference in the peak position of the shoulder and the one for the unstressed silicon and using the stress sensitivity as known from the literature [6, 15], results in tensile stresses in the order of 10 to 12 GPa, which is reasonable for a highly stressed zone around a crack tip and still lower than the theoretical tensile strength. As the signal intensity of the shoulder is low, the peak position cannot be determined with high accuracy. Therefore the calculation of the tensile stress is a rough estimation of the in-plane component of the stress tensor. As the spot size of the laser beam was 2 μm with a penetration depth of about 250 nm, the measured volume is bigger than the stress field around crack tips such as the one observed in Figs 5 and 6. This leads to the following conclusions: a) the intensity of the shoulder is small in comparison with the normal Raman peak, b) the lateral resolution of the microscope allows only measurement of an average tensile stress in the vicinity of the crack tip and c) cracks outside the laser focus spot were not detected. Taking into account the diameter of the laser focus spot as well as the penetration depth of the laser and comparing it with the cross-sectional area of the pillar, it is found that only 2% of the cross section area is covered by the laser. This explains why this behavior is observed only in 5% of the measurements. The fact that the shoulder appears only temporarily could also be explained with the laser spot size. As soon as the crack has propagated completely through the volume probed by the laser, the signal is no longer influenced by the crack since no stress exists at the crack surface.

2.6 Conclusions

In this contribution, a novel technique to characterize the mechanical properties of silicon micro-pillars combining micro-Raman spectroscopy with a micro-indenter has been presented. Tests can be performed at high data acquisition rates and also quickly after having automated the system. The main advantage of micro-Raman spectroscopy compared to other methods, including *in situ* SEM compression tests, is to not only gain information about the local compressive stresses, which can also be accessed during the experiment, but also locally to monitor certain microstructural details by the analysis of peak positions and shapes such as grain size, phase transformations and stress gradients. Verifying the type of microstructural changes by TEM analysis allows determining the point of occurrence of the microstructural changes. The method is applicable for specimen sizes from the millimeter to the sub-micrometer ranges. In the sub-micrometer range, the position accuracy of the laser with the optical microscope and the laser diameter are the only limitations.

It has been shown that the stresses derived from both the load cell and the stress-induced Raman shift agree very well with a discrepancy of less than 3%. In three experiments, a Raman band appeared at 490 - 520 cm^{-1} for compressive stresses ranging between -1.2 and - 2.5 GPa. Based on the TEM study, it was found that the main deformation mechanism of the pillars is brittle fracture. Dislocation nucleation and movement seems to play only a secondary role and to be triggered by interaction with cracks. Phase transformations could neither be observed during (by Raman spectroscopy) nor after the deformation (by TEM). Hence, we strongly believe that the shoulders observed are due to cracks propagating through the laser spots. This result is corroborated by comparing the theoretical stress of silicon (13 – 18.5 GPa) with the calculated in-plane tensile stress obtained from Raman shift (10 – 12 GPa).

Finally, the amalgamation of micro-Raman spectroscopy with a micro-indenter makes possible to locally detect micorstructural changes such as phase transformations and stress gradients induced by high compressive stress to silicon micro-pillars. The results demonstrate that the combination of both methods is an efficient technique not only to determine material properties in the micrometer range but also to identify the onset crack propagation as a potential solution for process control.

2.7 Acknowledgements

The authors would like to thank C. Grange for conducting the micro-compression tests, L. Barbieri for micro-pillar preparation, Joy Tharian for the FIB lamellea.

2.8 Appendix

This section presents data from uniaxial tensile tests performed on Ultra high molecular weight polyethylene (UHMWPE). These results are not directly correlated to the compression tests on silicon pillars. But the two studies show enough similarities in terms of method and results, that it makes sense to combine these two works. In both studies uniaxial stresses were applied to the samples and in both cases the measurements were performed *in situ*.

2.8.1 The Influence of Stress on the Peak Position of Polymers

So far we have described the influence of uniaxial stress on the peak position of silicon. But also the Raman peaks of polymers are sensitive to applied stress and show similar effects [28-31]. In this section the stress induced peak shift of polyethylene will be discussed.

Ultra high molecular weight polyethylene (UHMWPE) thin films with different draw ratios were tested. The polyethylene was pre-oriented and stretched at an elevated temperature of 135°C to draw ratios of $\lambda = 20$, 40 and 50. A detailed description of the process is reported elsewhere [32]. The draw ratio is defined as the ratio of the unit length after the deformation, L to and the initial unit length, L_0: $\lambda = L/L_0$. The drawing leads to a preferred orientation of the polyethylene chains. The degree of the orientation after the stretching process has an influence on the mechanical properties such as the Young's modulus and the strain at breaking. For example, Young's modulus ranges from 40, 70 to 80 GPa for $\lambda = 20$, 40 and 50. Figure 2.7 shows a section of a Raman spectrum of polyethylene. The peak at 1057 cm^{-1} belongs to an asymmetric stretching C-C vibration whereas the peak at 1123 cm^{-1} corresponds to a symmetric stretching C-C vibration. One should note that the resulting peak at 1123 cm^{-1} is the envelope of a "narrow" band and a "broad" band [30]. The peak at 1291 cm^{-1} and the peak at 1375 cm^{-1} belong to a twisting mode and a wagging mode of CH$_2$, respectively.

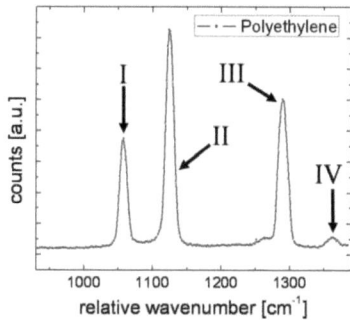

Figure 2.7 Section of a Raman spectrum of polyethylene. The arrows indicate to the Raman peaks: I: v_{as} (C-C) asymmetric stretching mode at 1057 cm^{-1}; II: v_s (C-C) symmetric stretching mode at 1123 cm^{-1}; III: v_t (CH$_2$) twisting mode at 1291 cm^{-1} and IV: ω(CH$_2$) wagging mode at 1375 cm^{-1}

To see the influence of applied strain on the Raman spectra of polyethylene stress-strain measurements were carried out using a tensile testing machine. The tensile machine was mounted below a confocal Raman microscope (WITec, CRM 200, wavelength: 442 nm) to measure the Raman signal *in situ*. The strain was applied parallel to the direction of the drawing process.

Figure 2.8 shows the Raman shift as a function of applied strain for the polyethylene film made with a draw ratio of 50. Three different peaks were analyzed. For the peak at 1123 cm^{-1}, only the narrow band was analyzed as the broad band at the position did not influence the result. From the figure, one can see that the applied strain leads to a shift to lower wavenumbers. Up to a strain of 2% the shift seems to be linear. For higher strains, the slopes of all three curves change. Furthermore, the three peaks show different strain dependencies. The peak at 1291 cm^{-1} is only altered slightly and has strain dependence of

0.16 cm^{-1}/%. In comparison the peaks at 1123 cm^{-1} and at 1057 cm^{-1} exhibit a strain dependence of 0.58 cm^{-1}/% and 0.75 cm^{-1}/%, respectively.

The reason for the different strain dependencies is the origin of the peaks. The peak at 1291 cm^{-1} belongs to a CH$_2$ wagging mode while the peak at 1057 cm^{-1} corresponds to a C-C molecule vibration, which is the backbone bond of the polyethylene. In a tensile experiment the stress is mainly applied on the back bond of the polymer. The C-H bonds of the polymer are like small side arms and are therefore much less affected by the stress.

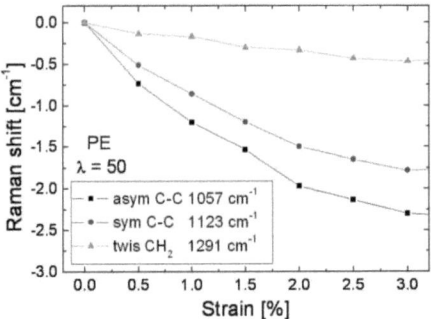

Figure 2.8 Raman peak shift as a function of applied uniaxial strain for a polyethylene film with a draw ratio of 50.

Figure 2.9 shows the peak shift for the peak at 1057 cm^{-1} for the draw ratios of $\lambda = 20$, 40 and 50 up to 5% strain. The peak corresponds to a C-C backbone vibration. All three samples show the same behavior. With increasing strain the peaks are shifted to lower wavenumbers, but the slope of the peak shift strongly depends on the draw ratio of the film. The higher the draw ratio, the bigger is the slope of the sample. Up to a strain of approximately 2.5% the slope is in all three samples linear. The sample with a draw ratio of 20 shows a linear behavior up to a strain of 5%. Above a strain of 2.5%, in the samples with a draw ratio of $\lambda = 40$ and 50 the peaks show no strain dependence. This could be due to plastic deformation or increased relaxations of the polymer films as the measurement were not performed *in situ*. Another reason could be creep of the samples at the clamping of the tensile machine.

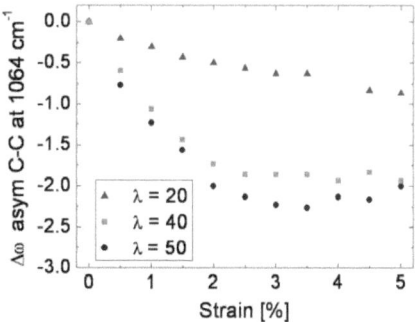

Figure 2.9 Raman peak shift of the peak at 1057 cm^{-1} as a function of the strain for the three different draw ratio λ = 20, 40 and 50.

Highly oriented polymers are assumed to consist of a number of anisotropic units with a uniform stress throughout the aggregate along the orientation axis [33]. Therefore, the macroscopic modulus can be expressed as:

$$\frac{1}{E} = \frac{1}{E_c} + \frac{\langle \sin^2 \theta \rangle}{G}, \quad (2.1)$$

where E is the macroscopic Young's modulus, E_c is the average chain modulus, G the macroscopic shear modulus, and $\langle \sin^2 \theta \rangle$ is the mean value of orientation over the chain in the units about orientation axis. Equation (2.1) indicates that the deformation in highly orientated polymer films can be divided into a crystal stretching part (elongation of the polymer chains) and a rotation (shear deformation of small domains containing the chain segment). For high draw ratios the angle θ becomes small and therefore the portion of the rotation in the equation (2.1) decreases. As a consequence, in highly orientated films applied strain or stress will strongly affect the elongation of the C-C backbone and therefore lead to a peak shift of the corresponding Raman peak. Maximal peak shifts could be obtained in polymer crystals where the chains are perfectly orientated and therefore θ becomes zero.

2.8.2 Acknowledgement (Appendix)

The authors would like to thank Marta Szankowska and Dario Corica for the examination of the Raman peak shifts related to uniaxial tensile stresses. Further we would like to thank Theo Thervoort for the supply of ultra high molecular weight polyethylene (UHMWPE) thin films.

2.9 References

1. F. Ericson and J.A. Schweitz, *Micromechanical fracture strength of silicon*. Journal of Applied Physics, 1990. **68**(11): p. 5840-5844.
2. O.M. Jadaan, N.N. Nemeth, J. Bagdahn, and W.N. Sharpe, *Probabilistic Weibull behavior and mechanical properties of MEMS brittle materials*. Journal of Materials Science, 2003. **38**(20): p. 4087-4113.
3. S. Greek, F. Ericson, S. Johansson, and J.A. Schweitz, *In situ tensile strength measurement and Weibull analysis of thick film and thin film micromachined polysilicon structures*. Thin Solid Films, 1997. **292**(1-2): p. 247-254.
4. T. Namazu, Y. Isono, and T. Tanaka, *Evaluation of size effect on mechanical properties of single crystal silicon by nanoscale bending test using AFM*. Journal of Microelectromechanical Systems, 2000. **9**(4): p. 450-459.
5. B. Moser, K. Wasmer, L. Barbieri, and J. Michler, *Strength and fracture of Si micropillars: A new scanning electron microscopy-based micro-compression test*. Journal of Materials Research, 2007. **22**(4): p. 1004-1011.
6. I. DeWolf, *Micro-Raman spectroscopy to study local mechanical stress in silicon integrated circuits*. Semiconductor Science and Technology, 1996. **11**(2): p. 139-154.
7. A.K.S. V.T. Srikar, M. Selim Ünlü, Bennett B. Goldberg, Mark Spearing, *Micro-Raman Measurement of Bending Stresses in Micromachined Silicon Flexures*. Journal of Microelectromechanical Systems, 2003. **Vol. 12**(No.6): p. 779-787.
8. A. Kailer, Y.G. Gogotsi, and K.G. Nickel, *Phase transformations of silicon caused by contact loading*. Journal of Applied Physics, 1997. **81**(7): p. 3057-3063.
9. Y. Gogotsi, T. Miletich, M. Gardner, and M. Rosenberg, *Microindentation device for in situ study of pressure-induced phase transformations*. Review of Scientific Instruments, 1999. **70**(12): p. 4612-4617.

10. Y. Gogotsi, G.H. Zhou, S.S. Ku, and S. Cetinkunt, *Raman microspectroscopy analysis of pressure-induced metallization in scratching of silicon*. Semiconductor Science and Technology, 2001. **16**(5): p. 345-352.
11. G.Y. Domnich V., *Phase Transformation in Silicon under Contact Loading*. Review of Advanced Material Science, 2002(3): p. 1-36.
12. R. Gassilloud, C. Ballif, P. Gasser, G. Buerki, and J. Michler, *Deformation mechanisms of silicon during nanoscratching*. Physica Status Solidi A-Applications and Materials Science, 2005. **202**(15): p. 2858-2869.
13. Z. Iqbal and S. Veprek, *Raman-scattering from hydrogenated microcrystalline and amorphous-silicon*. Journal of Physics C-Solid State Physics, 1982. **15**(2): p. 377-392.
14. J.K. Bhardwaj and H. Ashraf, *Advanced silicon etching using high density plasmas*. Micromachining and Microfabrication Process Technology, 1995. **2639**: p. 224-233.
15. R. Rabe, J.M. Breguet, P. Schwaller, S. Stauss, F.J. Haug, J. Patscheider, and J. Michler, *Observation of fracture and plastic deformation during indentation and scratching inside the scanning electron microscope*. Thin Solid Films, 2004. **469**: p. 206-213.
16. K.F. Dombrowski, *Micro-Raman investigation of mechanical stress in Si device structures and phonons in SiGe*. 2000, Brandenburgische Technische Universität Cottbus.
17. S. Rubanov and P.R. Munroe, *FIB-induced damage in silicon*. Journal of Microscopy-Oxford, 2004. **214**: p. 213-221.
18. L. Frey, C. Lehrer, and H. Ryssel, *Nanoscale effects in focused ion beam processing*. Applied Physics a-Materials Science & Processing, 2003. **76**(7): p. 1017-1023.
19. H. Zhang, B.E. Schuster, Q. Wei, and K.T. Ramesh, *The design of accurate micro-compression experiments*. Scripta Materialia, 2006. **54**(2): p. 181-186.
20. X.S. Zhao, Y.R. Ge, J. Schroeder, and P.D. Persans, *Carrier-induced strain effect in Si and GaAs nanocrystals*. Applied Physics Letters, 1994. **65**(16): p. 2033-2035.

21. J. Zi, H. Buscher, C. Falter, W. Ludwig, K.M. Zhang, and X.D. Xie, *Raman shifts in Si nanocrystals*. Applied Physics Letters, 1996. **69**(2): p. 200-202.
22. G. Weill, J.L. Mansot, G. Sagon, C. Carlone, and J.M. Besson, *Characterization of Si-III and Si-IV, metastable forms of silicon at ambient pressure*. Semiconductor Science and Technology, 1989. **4**(4): p. 280-282.
23. U. Dahmen, C.J. Hetherington, P. Pirouz, and K.H. Westmacott, *The formation of hexagonal silicon at twin intersections*. Scripta Metallurgica, 1989. **23**(2): p. 269-272.
24. S.J. Lloyd, J.M. Molina-Aldareguia, and W.J. Clegg, *Deformation under nanoindents in Si, Ge, and GaAs examined through transmission electron microscopy*. Journal of Materials Research, 2001. **16**(12): p. 3347-3350.
25. S.R. M. Levinshtein, M. Shur, *Handbook Series on Semiconductor Parameters, Vol. 1* ed. Ed. Vol. Vol. 1. 1996, Singapore: World Scientific.
26. S. Hoffmann, I. Utke, B. Moser, J. Michler, S.H. Christiansen, V. Schmidt, S. Senz, P. Werner, U. Gosele, and C. Ballif, *Measurement of the bending strength of vapor-liquid-solid grown silicon nanowires*. Nano Letters, 2006. **6**(4): p. 622-625.
27. R.D. Nyilas and R. Spolenak, *Orientation-dependent ductile-to-brittle transitions in nanostructured materials*. Acta Materialia, 2008: p. 5627-39.
28. K. Tashiro, G. Wu, and M. Kobayashi, *Morphological Effect On The Raman Frequency-Shift Induced By Tensile-Stress Applied To Crystalline Polyoxymethylene And Polyethylene - Spectroscopic Support For The Idea Of An Inhomogeneous Stress-Distribution In Polymer Material*. Polymer, 1988. **29**(10): p. 1768-1778.
29. J. Moonen, W.A.C. Roovers, R.J. Meier, and B.J. Kip, *Crystal and molecular deformation in strained high-performance polyethylene fibres studied by wide-angleX-ray scattering and Raman-spectroscopy*. Journal of Polymer Science Part B-Polymer Physics, 1992. **30**(4): p. 361-372.
30. W.F. Wong and R.J. Young, *Analysis of the Deformation of Gel-Spun Polyethylene Fibers Using Raman-Spectroscopy*. Journal of Materials Science, 1994. **29**(2): p. 510-519.

31. V.K. Mitra, W.M. Risen, and R.H. Baughman, *Laser Raman study of stress dependence of vibrational frequencies of a monocrystalline polydiacetylene.* Journal of Chemical Physics, 1977. **66**(6): p. 2731-2736.
32. J. Levèfre, *Ultra-high performance polymer foils*, in *Department of Materialscience*. 2008, ETH Zurich: Zurich. p. 35.
33. Y. Ward and R.J. Young, *Deformation studies of thermotropic aromatic copolyesters using NIR Raman spectroscopy.* Polymer, 2001. **42**(18): p. 7857-7863.

3 Measuring Stresses in Metal Thin Films by Means of Raman Microscopy Using Silicon as a Strain Gage Material

The second chapter showed the high accuracy of Raman spectroscopy for measuring one dimensional mechanical stresses. Pure uniaxial stresses are ideal for characterizing the properties of materials but are found seldom in application and devices. In this chapter we present the measurements of stresses in two dimensions. The new approach of this work is that stresses which were mapped in the silicon thin film by means of Raman microscopy were used to calculate stresses in an attached metal thin film. The experimental set-up required the crystallization of an amorphous sputtered silicon thin film in order to obtain a sufficient silicon Raman signal. This was accomplished by using the laser of the Raman microscope. The method of crystallization could also be applied for creating micro-sized crystalline silicon structures. The presented results have the character of a feasibility study. The approach used leads to reasonable and reproducible results. Nevertheless, a more evolved experimental set-up such as a using single crystal silicon thin film instead of a polycrystalline thin film would increase the accuracy of the data as well as the application possibilities. The illustrated results are accepted by the Journal of Raman spectroscopy:

Wermelinger T., Charpentier C., Yüksek M. D., Spolenak R.; *Measuring stresses in metal thin films by means of Raman microscopy using silicon as a strain gage material.* Aceppted by Journal of Raman spectroscopy

3.1 Abstract

Mechanical stresses in microelectronics and micro-electromechanical systems (MEMS) are usually much higher than in macroscopic counterparts. This is due to size effects in materials properties, i.e. yield strength, on one hand, and the joining of dissimilar materials with regards to the coefficient of thermal expansion, on the other. Furthermore high residual stresses can originate also from the deposition process utilized. These stresses and gradients thereof can lead to damage by fracture, creep and fatigue and thus pose a reliability issue. Identifying the locations of highest stresses in a device is crucial for reliability improvement. Currently, both Laue X-ray micro diffraction and convergent-beam electron diffraction are able to locally determine the stresses in metal thin films. Here, we propose a modified method of indirect Raman microspectroscopy to measure stresses with a lateral resolution in the submicrometer range at a laboratory scale. The method encompasses the crystallization of an amorphous silicon layer by local laser annealing and its subsequent usage as a strain gage. Stresses in an aluminum thin film were determined as a function of temperature. In the addition to the average stress, the stress distribution could also be monitored.

3.2 Introduction

The local measurement of mechanical stresses is motivated by reliability improvement as well as fundamental research. Firstly, it is known that stresses and stress gradients in micro-electromechanical systems (MEMS) can lead to failure by electromigration, fatigue, creep and fracture. Recently, Kahn *et al.* [1] revealed that the failure mechanism of cyclically loaded poly-Si MEMS is stress corrosion. Mechanical stresses can originate from differences in the coefficients of thermal expansion of the different materials. Consequently, cyclic temperature changes have been shown to lead to fatigue [2]. Other causes of intrinsic stresses are stresses induced during thin film deposition, for example, the formation of chemical vapor deposited silicon nitride films. In this case, the occurring stresses can be adjusted by changing the deposition temperature as well as the stoichiometry of the gases [3]. Large stresses can also be induced in the substrate near embedded structures such as trenches which can cause residual stresses or stresses can be applied externally during operation and deployment [4-6]. The problems associated with these stresses are various. One of the most serious issues is a stress-induced trigger of nucleation and propagation of dislocations and the formation of voids and cracks, which can lead to short-circuits and failure. To know why and where high stresses occur with a high as possible lateral resolution could help improving the reliability of micro-electromechanical systems.

Secondly, micro-X-ray studies have shown that stresses observed in thin metal films are strongly non-uniform [7-9]. Interactions between neighboring grains lead to highly non-uniform stresses in an individual grain. It was found that grains with diameters much bigger than the film thickness deform in a highly inhomogeneous way. For a better understanding of mechanical behavior of thin metal films a method for stress measurements with a high lateral resolution has to be available.

Several techniques are commonly used for stress measurements but none of them is without shortcomings when applied to materials used in microelectronics. One method is micro-X-ray diffraction with a spot size of less than one micron in diameter [10]. The drawback of this method is that it is only available at a few sources worldwide. Another technique to investigate thin film plasticity is convergent-beam electron diffraction [11]. The lateral resolution is very high (only a few nm) but the samples must be thinned to electron transparency. Therefore the stress state under investigation is usually relaxed or altered in comparison to the original stress state.

Another technique to measure stress is Raman spectroscopy. Since the first reports of Anastassakis *et al*. [12], it is known that Raman peaks are sensitive to stresses. Ossikovski *et al*. [13] showed that it is possible to determine all six components of the silicon strain tensor by off-axis illumination and polarization of the incident and scattered light. Bonera *et al*. [14] presented another method to measure three components of the strain tensor by using a micro-Raman microscope. The lateral resolution is defined by the optics and the wavelength of the laser light, and can be in the submicrometer range [15]. Ma *et al.* applied the method to measure thermal stress in metallic interconnects by measuring the peak shift in the silicon substrate [16]. Furthermore, from the Raman spectra the information about the microstructure such as the phase or the grain size and the temperature is also accessible [17-20]. In transparent materials, a confocal microscope allows the mapping of peak shifts in 3-D [21].

This paper presents a method for measuring stresses in thin aluminum films using a silicon thin film as a stress gage. Amorphous silicon was sputtered onto a silicon nitride membrane. Laser irradiation was used to crystallize the silicon thin film. On the opposite side of the silicon nitride film an aluminum film was sputtered. Thermal stresses were induced as a poly-Si silicon nitride aluminum multilayer structure by heating to 200°C.

3.3 Experimental

A silicon thin film with a thickness of approximately 280 nm was prepared by DC magnetron sputtering using a PVD Products™ sputtering device. The sputtering was performed in an argon atmosphere at a pressure of $1.5 * 10^{-7}$ Torr and at room temperature. The discharge power was held at 500 W. The silicon was sputtered on a membrane optimized for TEM observation which consisted of a 500x500 μm^2 large silicon nitride (SiN_x) membrane with a thickness of 30 nm. Figure 3.1 a) shows a schematic cross section of the membrane.

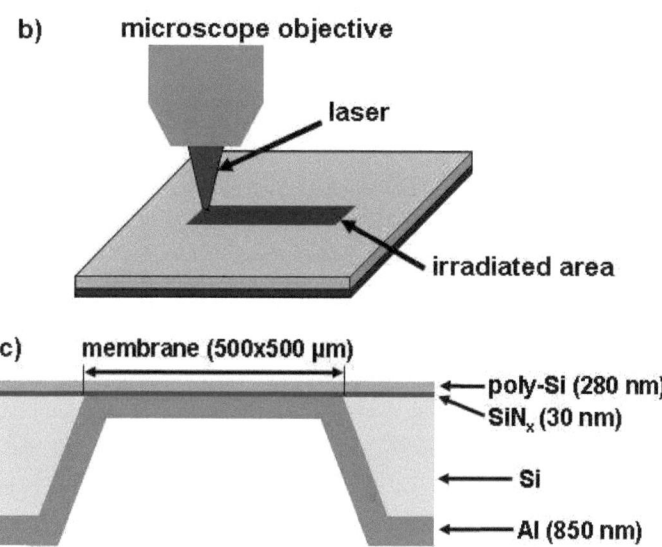

Figure 3.1 a) Schematic cross section of the 280 nm silicon thin film sputtered on a TEM sample holder consisting of a 30 nm thick and 500x500 μm^2 wide membrane. b) Laser irradiation of the silicon thin film using a confocal Raman microscope. c) 850 nm aluminum was sputtered on the back side of the TEM sample holder.

Defined areas of the sputtered silicon thin films were irradiated with a laser using a confocal Raman microscope (CRM 200, WITec) (Figure 3.1 b)). The continuous He-Cd-laser had a wavelength of 442 nm and a power of about 1.6-1.8 mW. The laser had a diameter of 300 nm which led to a maximal power density of around $5.6-6.3*10^5$ W/cm^2. The power density of the laser was adjustable. The laser irradiations were performed at ambient conditions with different intensities and different scan speeds in the range of 3 to 10 μm/sec. The energy density was calculated with the scan speed and is given by: $E = p*t$, whereas p is the power density and t is the time for a certain point staying within the laser focus. This time depends only on the scan speed and the laser diameter.

After the laser irradiation, 850 nm of aluminum were sputtered on the backside of the membrane (see Figure 3.1 c). Sputtering was performed at a pressure of $3.5*10^{-7}$ Torr at room temperature. The applied discharge power was set at 280 W. The silicon-silicon nitride-aluminum structured was annealed for 4 hours at 400 °C.

A FEI TEM (transmission electron microscope) CM30 working at 300 kV and a confocal Raman microscope were used to analyze the microstructure of the silicon thin film before and after the laser irradiation. The microstructure of the aluminum thin film was examined by means of TEM before and after the heating treatment. The average grain sizes of the silicon and aluminum films were evaluated using the academically developed image analysis software "LINCE" [22].

The multilayer structure was step-wise heated up to maximal 200°C to induce mechanical stresses due to the mismatch of the thermal expansion coefficients of silicon and aluminum. The stresses were mapped with the confocal Raman microscope in the area where laser irradiation was performed. To prevent the Raman laser from inducing significant heating effects and therefore influencing the measurements, the intensity of the laser was lowered to approximately 5% of the maximal power.

3.4 Results

3.4.1 Silicon Thin Film

Figure 3.2 a) and b) show TEM images of the silicon thin film as-sputtered and after laser irradiation with the He-Cd laser, respectively. In Figure 3.2 a), the grain size of the as-sputtered silicon is not detectable. The diffraction pattern of the area, visible as an inlay, shows broad and continuous rings. This indicates that the sputtered silicon is amorphous. Figure 3.2 b) shows the silicon thin film after laser irradiation with an energy density of $5*10^4$ J/cm^2. The grains have grown and exhibit an average grain size of 35 ± 17 nm. The according diffraction pattern of the area shows brightly focused spots in addition to the continuous rings. The spots clearly indicate the polycrystalline character of the film. The presence of the still continuous rings indicates that the film either remains amorphous to a small degree or some very small grains were detected. The white circle marks a region with moiré fringes, which means that the grains do not have a columnar structure. Higher energy densities of $1*10^5$ J/cm^2 lead to relatively large single crystals (see Figure 3.2 c)). The diffraction pattern, taken from the area within the white circle, reveals the single-crystalline character of grain.

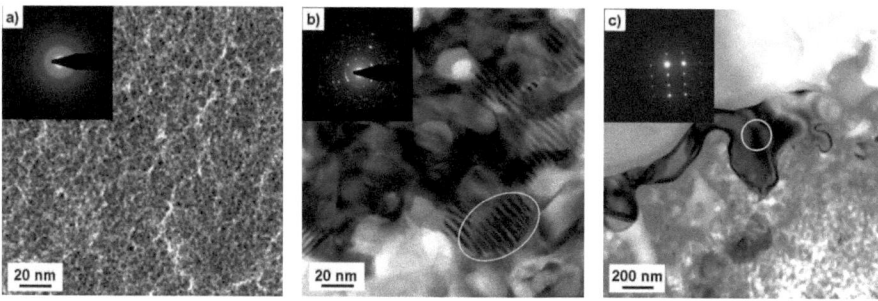

Figure 3.2 a) TEM image of the silicon thin film as sputtered by the magnetron sputtering tool. The inlay shows the diffraction pattern of the image. The broad rings belong to amorphous silicon. b) TEM image of the silicon thin film after laser irradiation with an energy density of $5*10^4$ J/cm^2. Grains are visible. The white circle points out a moiré fringe, and therefore a non-columnar structure. The diffraction pattern

shows bright spots belonging to the polycrystalline material. c) TEM image of the silicon thin film after laser irradiation with an energy density of $1*10^5$ J/cm². The white circle marks the area where a diffraction pattern was taken. This regions is clearly single crystalline.

The transition from the amorphous region to the crystalline region is well defined. Figure 3.3 shows two images of laser irradiated areas taken with TEM and optical microscopy. The TEM image (Figure 3.3 a)) illustrates that the grain growth can be performed very selectively in specific areas. The polycrystalline line has a diameter of 1.5 µm. The optical microscope image (Figure 3.3 b)) shows in addition to the amorphous area and the irradiated line also a third area which is surrounding the polycrystalline area. The black line refers to the location where a Raman line scan was performed.

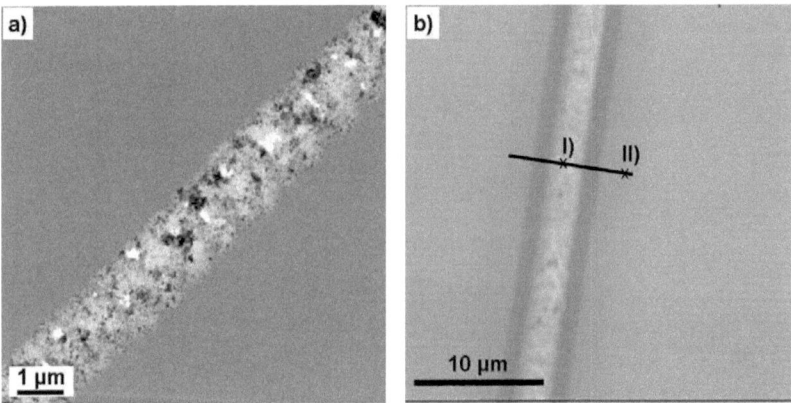

Figure 3.3 a) TEM image of a 1.5 µm wide polycrystalline line surrounded by amorphous silicon. b) Optical microscope image of a similar structure with the location of a Raman line scan indicated in black.

Figure 3.4 a) shows the results of the Raman scan performed across the laser irradiated area. Plotted is the peak position as a function of the measured distance along the scan line. In the as-sputtered region the silicon Raman peak is from approximately 450 to 458 cm⁻¹. Near the irradiated area the peak position is strongly increasing. Approaching the irradiated area, the peak position shifts upwards to 518 cm^{-1} within a 1.5 µm span. Once within the irradiated area, the peak position is very uniform. In Figure 3.4 b) two Raman spectra from the line scan are extracted (as indicated in Figure 3.3). Spectrum I) was measured in the laser irradiated area. It exhibits a crystalline spectrum. Spectrum II) was collected on the as-sputtered area. Instead of a sharp peak only a broad peak at approximately 458 cm^{-1} is visible. Such a spectrum corresponds to amorphous silicon.

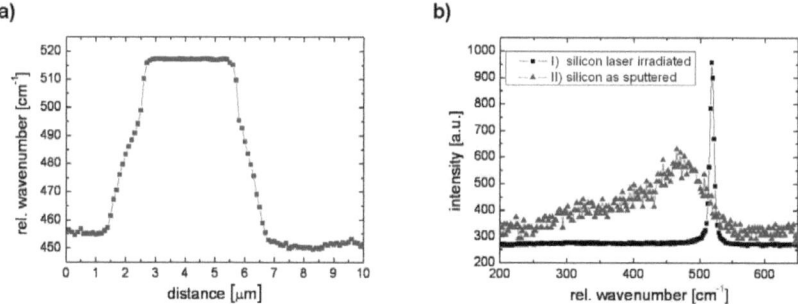

Figure 3.4 Results of the Raman line scan as it can be seen in Figure 3.3 b). a) The peak position of the silicon Raman peak as a function of the distance along the scan line. b) A comparison of a spectrum taken from the middle of the irradiated area, I), and a spectrum taken from the as-sputtered part of the silicon thin film, II).

Adjusting the density of scanned lines per area as well as the scan speed it was possible to produce homogeneous, micrometer-sized areas consisting of poly-Si in the otherwise amorphous membrane. Figure 3.5 shows the optical microscope image of an irradiation-produced 5x10 µm^2 polycrystalline area. This region was used to map the peak shift during the heating experiments.

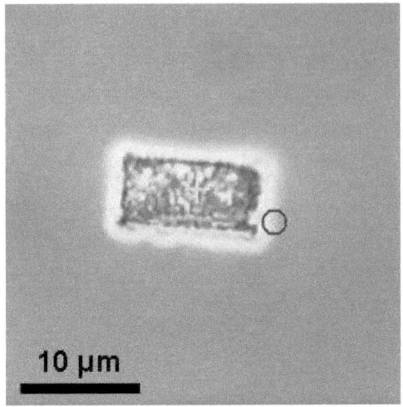

Figure 3.5 Optical microscope image of a 5x10 μm^2 irradiated area. The area was used to measure the peak shift during the heating experiments up to 200°C.

3.4.2 Aluminum Thin Film

Figure 3.6 shows TEM images of the aluminum thin film as-sputtered and after annealing for 4 hours at 400°C. The annealing caused grain growth. While the sputtered film had an average grain size of 94 ± 20 nm, the grain size increased during the annealing to 226 ± 67 nm. Nevertheless, the grain size remains small in comparison to the film thickness of 850 nm. Although no moiré fringes are visible, the differences in contrast within certain grains could indicate that the grain structure is non-columnar.

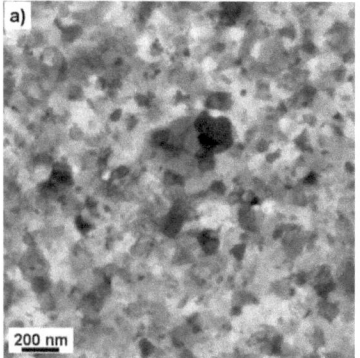 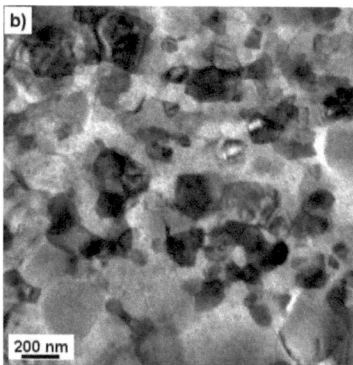

Figure 3.6 a) TEM image of the as-sputtered aluminum thin film. b) TEM image of the aluminum thin film after annealing for 4 hours at 400°C.

3.5 Discussion

3.5.1 Grain Growth in Silicon Thin Films

The TEM images in Figure 3.3 as well as the Raman experiments (see Figure 3.4) exhibit that laser irradiation leads to significant grain growth. Many studies of laser induced grain growth in silicon thin films on bulk substrates showed that the energy density is on the order of 0.5 J/cm^2 [23-27]. Choi *et al.* [28, 29] used a bridge structure which was very similar to the experimental set-up presented in this paper (see Figure 3.1a)) to induce grain growth in a 40 nm thick silicon film. In their case, the energy density was in the range of 0.17-0.3 J/cm^2. Surprisingly, in our experiments we found an energy threshold for grain growth of 500 J/cm^2. Thus let us consider in detail, how energy densities were determined. The energy density for grain growth was calculated using the scan speed and the power density of the laser. The calculated energy density values for our experiments are rough estimations because the diameter of the focused laser, which influences the power density, changes during the experiment when the sample moves out of focus. The energy densities used in the presented experiments were at least three orders of magnitude higher than the values stated in literature. The main difference between the set-ups found in literature and our set-up was the wavelength of the laser. Here a continuous laser with a wavelength of 442 nm was used, while in literature the lasers were used in a pulsed mode with wavelengths in the UV range, varying from 193 nm to 351 nm. The optical absorption coefficient of silicon drastically increases in the UV range and is up to two magnitudes higher than at a wavelength of 442 nm [30]. Another major difference is the spot size of the laser. This work's continuous laser had a diameter of 300 nm. Therefore only a small spot was heated and the sample had to be scanned to produce crystalline areas. A small diameter allows for selective grain growth only in specific areas and produces structures with a diameter in the micrometer range (see Figure 3.3 a)).

3.5.2 Grain Size

The average grain size of the silicon film in the irradiated area is 35 ± 17 nm, which falls below the lateral resolution of the microscope. Thus, several grains are always within the laser spot and give response to the measured signal. It is not possible to measure differences in the stress state of single silicon grains of this size.

The annealing led to an increase in the average size of the aluminum grains. The grain size changed from 94 ± 20 nm in the as-sputtered state to 226 ± 67 nm after the heat treatment. Having aluminum grains with sizes significantly larger than the lateral resolution the silicon thin film could be useful as a strain gage to map stresses of individual grains. However, in both cases the grain size stayed below the diameter of the laser spot. Still, it is possible to analyze the general stress state.

3.5.3 Stress Measurement

Figure 3.7 illustrates the Raman peak position of the poly-Si area (indicated in Figure 3.5) for room temperature, 70°C, 140°C and 200°C. The changing scale bars of the figures demonstrate that increasing temperatures lead to general shifts of the peak position to lower wavenumbers. In all mappings the poly-Si region was surrounded by amorphous silicon exhibiting a peak position of 460 to 470 cm^{-1}. Therefore, this region appears black in all figures. The peak shift consists of two different components. Firstly, the peak position depends directly on the temperature of the sample [20, 31, 32]. Secondly, the mismatch of the thermal expansion coefficients of silicon and aluminum induces thermal stresses into the multilayer structure. These stresses lead to shifts of the Raman peak position.

In the four scans of Figure 3.7 the range of the scale bar is 1.5 cm^{-1}. In contrast to the scan performed at room temperature the scans at elevated temperatures show a similar pattern in terms of regions with slightly higher or lower wavenumbers. The "room temperature" map has a less pronounced pattern than the mappings performed at higher temperatures. The temperature is homogeneous over the whole sample and therefore should lead only to a homogenous shift to lower wavenumbers. Small changes of the aluminum-silicon-film thickness-ratio could lead to inhomogeneous stresses in the multilayer structure.

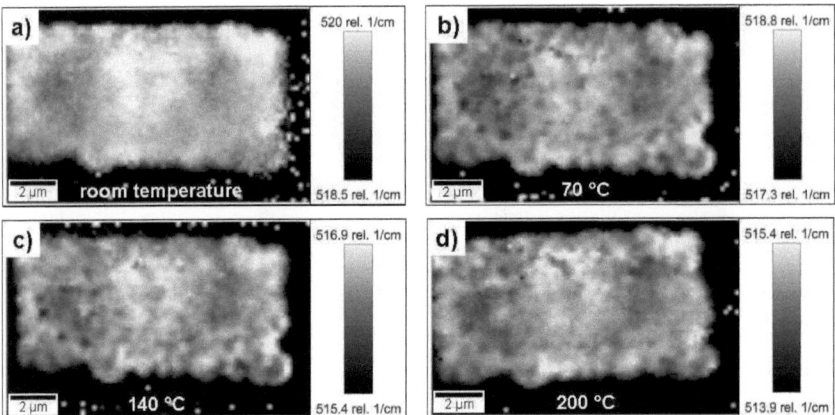

Figure 3.7 Raman maps of the poly-Si area in the as-sputtered state (see Figure 3.5) at room temperature, 70°C, 140°C and 200°C. The silicon Raman peak decreases with increasing temperature due to thermal stresses and temperature.

To be able to measure the appearing stresses, the thermal and the stress related effects have to be separated. The approach used was to average all the spectra from the irradiated area.

Figure 3.8 a) illustrates the absolute peak shift of the averaged spectra of a silicon wafer, two poly-Si thin film areas produced by means of laser irradiation as well as the theoretical peak shift of silicon [20] as a function of the temperature. All the samples were step-wise heated up to 200°C. Moreover, "polycrystalline area_1" was cooled down afterwards to room temperature again. The temperature dependence of the poly-Si areas was highly similar to the one of single crystalline silicon and the theoretical calculated one. The maximal difference between the polycrystalline samples and the silicon wafer was 0.2 cm^{-1}.

Figure 3.8 b) shows the absolute averaged peak shift of always the same poly-Si area of the multilayer structure (see Figure 3.7) as a function of temperature in comparison with the peak shift of a silicon wafer. The first experiments were performed in the as-sputtered state of the. After an annealing. "aft. anneal. cycle1" was step-wise heated up to 200°C while in a second measurement (indicated as "aft. anneal. cycle2") also during the cooling maps were acquired. The silicon Raman peaks from the multilayer structures exhibited significantly stronger peak shifts than the silicon wafer. "aft. anneal. cycle1" displayed with a shift of 5.5 cm^{-1} at 200°C the largest temperature dependency. The two other measurements show a similar behavior and are shifted at 200°C by 4.4 to 4.6 cm^{-1} while the Si-wafer is only shifted by 3.8 cm^{-1}. The slope of "aft. anneal. cycle2" showed the same behavior during the heating-up and the cooling-down. No hysteresis was visible. Figure 3.8 a) showed that the temperature dependence of poly-Si was very similar to the one of the silicon wafer. Therefore, the additional peak shift the poly-Si on the multilayer structure had to be due to thermal stress origin from the mismatch of the thermal expansion coefficient of the different layers.

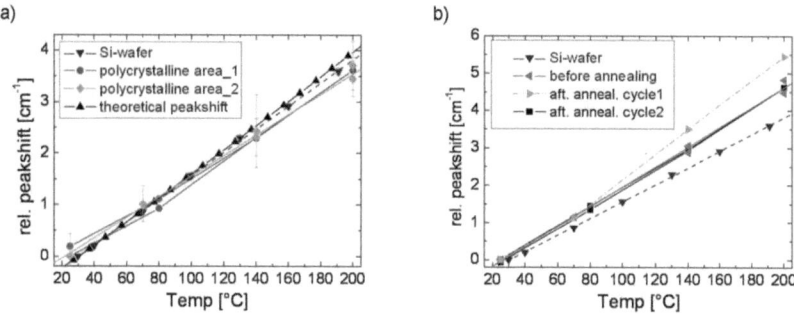

Figure 3.8 Absolute peak shift of the averaged areas with increasing temperature. a) Results of a silicon wafer, two laser irradiated poly-Si areas and the theoretical peak shift of silicon [20]. b) The temperature dependence of the same poly-Si area on the multilayer structure before annealing and two times after the annealing. The measurements were compared to the results from the Si-wafer.

The error bars in Figure 3.8 a) and b) derive form the standard deviations of the peak positions of the averaged areas. The Si-wafer was single crystalline and had in comparison with all measurements performed on poly-Si a much lower standard deviation (see Figure 3.9). But all samples showed an increased standard deviation with increasing temperature. The increases of the standard deviation in "before annealing" and "after anneal. cycle2" were in the order of 0.1 to 0.16 cm^{-1}. The standard deviation of the Si-wafer and "after anneal. cycle1" only increased 0.025 cm^{-1}. The stronger increase of the deviation of the multilayer structures could indicate that the stress distribution at elevated temperatures was laterally not homogenous. "polycrystalline area_2" showed a much higher standard deviation of 0.35 to 0.7 cm^{-1}. It is assumed that the large standard deviation was caused by very inhomogeneous grain growth.

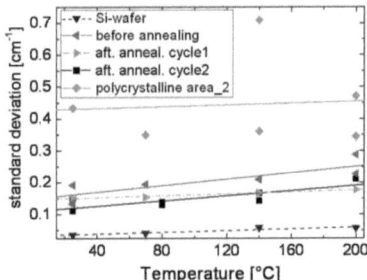

Figure 3.9 Standard deviation as a function of the temperature of a Si-wafer, the multilayer structure ("before annealing", "aft. anneal. cycle1" and aft. anneal. cycle2") and a poly-Si area. Higher temperatures led to an increased standard deviation. The multilayer structure showed stronger temperature dependence than a Si-wafer. The scattering of the poly-Si area was drastically higher due to an inhomogeneous grain growth.

At 200°C, the difference between the shifts of the Si-wafer and "aft. anneal. cycle1" was 1.66 cm^{-1}. In the two other measurement the difference was between 0.7 and 0.9 cm^{-1}. To calculate mechanical stresses from these differences, the crystal orientation of the grains is important as it influences the stress dependency of the peak shift [33]. It is known [34-37] that laser irradiated poly-Si thin films demonstrate a strong (111)-orientation. The structure of the samples allowed reducing the stress tensor to a biaxial isotropic stress which has the following simple form:

$$\sigma = \begin{pmatrix} \sigma_{xx} & 0 & 0 \\ 0 & \sigma_{yy} & 0 \\ 0 & 0 & 0 \end{pmatrix}; \text{ with } \sigma_{xx} = \sigma_{yy}. \quad (3.1)$$

Broblik [38] calculated the stress dependency of the peak shift for isotropic biaxial stresses of (111)-oriented silicon (see Table 1 in ref. [38]). As the Raman measurements were performed in a backscattering mode the silicon Raman signal consisted of a singlet and a doublet mode [33]. Both modes have the same peak position in the stress-free state, but they do not show the same stress dependency. The resolution of the spectrometer used was too low to distinguish the two different peaks. Therefore, the from the peak shifts calculated stresses were the mean values of the singlet and the doublet. Different values exist for the phonon deformation potential as it is determined from experiments. Figure 3.10 shows the stresses in the multilayer structure calculated by using the phonon deformation potential of Chandrasekhar *et al.* [39]. For comparison, in one case the phonon deformation potential of Anastassakis *et al.* [33] was used ("aft. anneal. cycle1 ana").

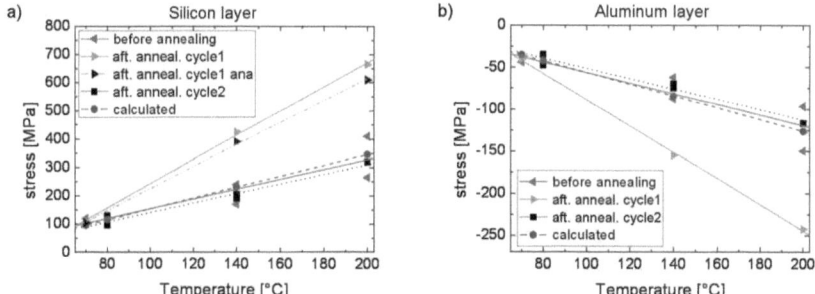

Figure 3.10 Stresses in the a) silicon layer and in the b) aluminum layer calculated from the peak shifts of the poly-Si area of the multilayer structure at different temperatures using the phonon deformation potential of Chandrasekhar *et al.* [39] and in case of "aft. anneal. cycle1 ana" Anastassakis *et al.* [33]. The curve "calculated" corresponds to the theoretical thermal stress in the silicon and the aluminum layer, respectively.

The stresses measured by means of Raman microscopy were compared to theoretical calculated values considering the different thermal expansion coefficients of the different materials. The multilayer structure has to fulfill the following assumption:

$$\sum f_i = 0 = \sum \sigma_i \cdot d_i \text{, with } f_i = \varepsilon_i \cdot E_i \cdot d_i, \qquad (3.2)$$

whereas f_i is the applied force/length of a layer, σ_i is the stress and d_i is the thickness of the *i-th* layer. The stress is defined by the strain ε_i and the Young's modulus E_i. The strain of a certain layer (e.g. silicon) is given from the difference between the thermal expansion of a freestanding silicon film and the thermal expansion of the silicon in the multilayer structure where its expansion is restricted by the other layers. This equation is valid if the whole multilayer structure is in the tensile stress regime and the radius of curvature is negligible. The validity of the assumptions is supported by literature [40-43] as well as by the finding that neither in optical microscope images nor in SEM images a significant curvature of the crystalline area was detected. The strain ε_i is defined as:

$$\varepsilon_i = \frac{l_i - l_{multilayer}}{l_i}. \qquad (3.3)$$

$l_{multilayer}$ is the thermal expansion of the multilayer structure. l_i is the thermal expansion of a freestanding layer and is given by:

$$l_i = l_0(1 + \Delta T \alpha_i). \qquad (3.4)$$

In the equation l_0 is the starting length, ΔT the temperature difference and α_i the thermal expansion coefficient. These equations allowed calculating the stress of every single layer of the multilayer structure.

The thermal expansion coefficient of poly-Si is $2.9*10^{-6}$ /K [44]. The thermal expansion of aluminum between 25°C and 200°C is in approximately 23 to $27*10^{-6}$ /K [45]. For simplicity the silicon nitride layer was added to the layer of poly-Si as the thermal expansion coefficient of silicon nitride [46] as well as the Young's modulus of both materials are comparable [47]. For the calculation the multilayer structure was reduced to a binary system consisting of 850 nm aluminum and 310 nm poly-Si.

Figure 3.10 indicates that with the exception of "aft. anneal. cycle1" the calculated stress and the measured stresses agree well with each other. Using the phonon deformation potential of Chandrasekhar led to slightly higher stress values in comparison with the one of Anastassaskis. The difference between "aft. anneal. cycle1" and "aft. anneal. cycle1 ana" at 200°C was 55 MPa which corresponds to a discrepancy of 7%. Taking into account the standard deviation of the peak position the differences are small. Therefore, in the following sections only the values of Chandrasekhar will be used.

The stress in the silicon layer was in the tensile state. To fulfill the requirements of (3.2 the aluminum layer had to be in the compressive state. The maximal stresses in the poly-Si film were at 200°C for "before annealing" and "aft. anneal. cycle2" approximately 250 MPa to 440 MPa. Deduced from the maximal stresses in the poly-Si film the maximal compressive stresses in the aluminum in the two measurements were about -90 MPa to -160 MPa. This amount of compressive stress could have induced plastic deformation [48, 49]. Plastic deformation would lead to a hysteresis during a heating up-cooling-down cycle such as was performed in "aft. anneal. cycle2". "aft. anneal. cycle2" did not exhibit such a behavior and no hysteresis was visible. Therefore, one can conclude that the aluminum film was always in the elastic regime.

Measurement "aft. anneal. cycle1" showed high tensile stresses of about 610 MPa to 660 MPa in the silicon film which correspond in the aluminum film to compressive stresses of -220 to -240 MPa. According to literature [48, 49], these stress values would have been above the yield strength of aluminum thin films. Plastic deformation would directly lead to a decrease of the slope. "aft. anneal. cycle1" did not show such a behavior. A plausible explanation could be that the laser intensity during the measurements was not stable. Figure 3.11 demonstrates the strong dependence of the peak position of poly-Si thin films from the intensity of the laser as the laser can strongly heat up a sample. Up to a normalized intensity of 0.7 (corresponds to 1.1-1.3 mW) the curves show a linear behavior with a slope ranging from -3.65 to 4.25 cm^{-1} per 0.1 Δ-intensity. Between a normalized intensity of 0.7 and the maximal intensity (1.6-1.8 mW) the curves differ strongly in terms of the peak position and behavior. But in all curves the slope decreases strongly.

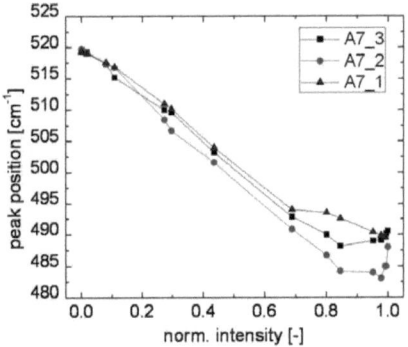

Figure 3.11 Peak position of a poly-Si thin film as a function of the normalized intensity of the Raman laser which had a maximal power of 1.6-1.8 mW.

To minimize heating effects caused by the laser, the intensity in all measurements was kept at about 5% of the maximal intensity. "aft. anneal. cycle1" is at 200° shifted by 0.6-1 cm^{-1} more than "before annealing" and "aft. anneal. cycle2". In the linear region of the curves such a shift is caused by a change of the laser intensity of 1.5 to 2.7%. As the intensity was adjusted manually such a deviation seems to be highly plausible.

3.6 Conclusions

A multilayer structure consisting of 850 nm Al, 30 nm amorphous SiN and 280 nm amorphous silicon was produced. Grain growth was induced by means of laser irradiation in particular areas of the amorphous silicon film. The produced Si- grains had an average diameter of 35 ± 17 nm and were sufficiently large to obtain a crystalline Raman signal. The small diameter of the laser allowed grain growth in highly specific areas and producing crystalline structures with a size in the micrometer range. As the crystalline silicon has other material properties than the amorphous silicon such as etching rates or electrical conductivity this could be a useful tool to produce freestanding structures or electrical conductors.

The multilayer structure was several times heated up to 200°C. The peak shift of the silicon Raman signal was mapped always in the same 5x10 μm^2 large crystalline region. The peak shift is shifted to lower wavenumbers due to a thermal and a stress induced component. Appearing inhomogeneous peak shifts are addressed to laterally inhomogeneous stresses. Nevertheless, such inhomogeneties are only visible to a small extent (0.18-0.23 cm^{-1}).

From the difference between the purely thermally induced peak shift of a silicon wafer and the measured peak shift of the multilayer structure the thermal stresses in the different layers were calculated. "before annealing" and "aft. anneal. cycle2" exhibited in the silicon layer a maximal thermal stresses of 250 to 450 MPa and are in good agreement with the theoretical calculated stress of 350 MPa. "aft. anneal. cycle1" showed tensile stresses up to 660 MPa. "before annealing" and "aft. anneal. cycle2" exhibited maximal compressive stresses of -90 to -160 MPa. These values are in the elastic range., In "aft. anneal. cycle1" most probably a slightly increased laser intensity caused local heating of the sample and therefore influenced the accuracy of the measurement. We have to point out that in the present study only thermal stresses were measured. Preexisting stresses due to the sputtering, laser irradiation or thermal treatments were not analyzed.

In summary the main points of this paper are:

- The local crystallization of amorphous Si thin films to potentially be used for micropatterning.
- The establishment of a method of using Raman microscopy to indirectly measuring stresses in thin metal films, which are non Raman active materials. This constitutes the first step toward local stress determination in such films, where the resolution of the Raman method could be enhanced by tip enhanced Raman spectroscopy.

3.7 Acknowledgement

The authors thank Christian Solenthaler, Susanne Köppl and Giancarlo Pigozzi for the TEM analysis as well as the helpful discussions. The authors acknowledge support of the Electron Microscopy Centre of the Swiss Institute of Technology (EMEZ). Further, Cesare Borgia and Noble Woo are acknowledged for their assistance with the magnetron sputtering tool. This work was supported by the ETH Research Grant TH -39/05-1.

3.8 References

1. D.F. Bahr, B.T. Crozier, C.D. Richards, and R.F. Richards, *Fatigue and fracture in membranes for MEMS power generation.* Mechanical Properties of Structural Films, 2001. **1413**: p. 28-36.
2. J.R. Lloyd, *Electromigration in thin film conductors.* Semiconductor Science and Technology, 1997. **12**(10): p. 1177-1185.
3. M. Pletea, W. Bruckner, H. Wendrock, and R. Kaltofen, *Stress evolution during and after sputter deposition of Cu thin films onto Si(100) substrates under various sputtering pressures.* Journal of Applied Physics, 2005. **97**.
4. I. DeWolf, *Micor-Raman spectroscopy to study local mechanical stress in silicon integrated circuits.* Semiconductor Science and Technology 1995. **11**: p. 139-154.
5. I. Dewolf, J. Vanhellemont, A. Romanorodriguez, H. Norstrom, and H.E. Maes, *Micro-Raman Study Of Stress-Distribution In Local Isolation Structures And Correlation With Transmission Electron-Microscopy.* Journal of Applied Physics, 1992. **71**(2): p. 898-906.
6. A.K.S. V.T. Srikar, M. Selim Ünlü, Bennett B. Goldberg, Mark Spearing, *Micro-Raman Measurement of Bending Stresses in Micromachined Silicon Flexures.* Journal of Microelectromechanical Systems, 2003. **Vol. 12**(No.6): p. 779-787.
7. R. Spolenak, W.L. Brown, N. Tamura, A.A. MacDowell, R.S. Celestre, H.A. Padmore, B. Valek, J.C. Bravman, T. Marieb, H. Fujimoto, B.W. Batterman, and J.R. Patel, *Local plasticity of Al thin films as revealed by X-ray microdiffraction.* Physical Review Letters, 2003. **90**(9).
8. M.A. Phillips, R. Spolenak, N. Tamura, W.L. Brown, A.A. MacDowell, R.S. Celestre, H.A. Padmore, B.W. Batterman, E. Arzt, and J.R. Patel, *X-ray microdiffraction: local stress distributions in polycrystalline and epitaxial thin films.* Microelectronic Engineering, 2004. **75**(1): p. 117-126.
9. R. Spolenak, O. Kraft, and E. Arzt, *Effects of alloying elements on electromigration.* Microelectronics Reliability, 1998. **38**(6-8): p. 1015-1020.

10. N. Tamura, A.A. MacDowell, R. Spolenak, B.C. Valek, J.C. Bravman, W.L. Brown, R.S. Celestre, H.A. Padmore, B.W. Batterman, and J.R. Patel, *Scanning X-ray microdiffraction with submicrometer white beam for strain/stress and orientation mapping in thin films*. Journal of Synchrotron Radiation, 2003. **10**: p. 137-143.
11. S. Kramer, J. Mayer, C. Witt, A. Weickenmeier, and M. Ruhle, *Analysis of local strain in aluminium interconnects by energy filtered CBED*. Ultramicroscopy, 2000. **81**(3-4): p. 245-262.
12. V. Senez, A. Armigliato, I. De Wolf, G. Carnevale, R. Balboni, S. Frabboni, and A. Benedetti, *Strain determination in silicon microstructures by combined convergent beam electron diffraction, process simulation, and micro-Raman spectroscopy*. Journal of Applied Physics, 2003. **94**(9): p. 5574-5583.
13. Anastass.E and E. Burstein, *Morphic Effects .1. Effects of External Forces On Photon-Optical Phonon Interactions*. Journal of Physics and Chemistry Of Solids, 1971. **32**(2): p. 313-&.
14. R. Ossikovski, Q. Nguyen, G. Picardi, J. Schreiber, and P. Morin, *Theory and experiment of large numerical aperture objective Raman microscopy: application to the stress-tensor determination in strained cubic materials*. Journal of Raman Spectroscopy, 2008. **39**(5): p. 661-672.
15. E. Bonera, M. Fanciulli, and D.N. Batchelder, *Combining high resolution and tensorial analysis in Raman stress measurements of silicon*. Journal of Applied Physics, 2003. **94**(4): p. 2729-2740.
16. B. Dietrich, V. Bukalo, A. Fischer, K.F. Dombrowski, E. Bugiel, B. Kuck, and H.H. Richter, *Raman-spectroscopic determination of inhomogeneous stress in submicron silicon devices*. Applied Physics Letters, 2003. **82**(8): p. 1176-1178.
17. Q. Ma, S. Chiras, D.R. Clarke, and Z. Suo, *High-Resolution Determination Of The Stress In Individual Interconnect Lines And The Variation Due To Electromigration*. Journal of Applied Physics, 1995. **78**(3): p. 1614-1622.
18. K. Wasmer, T. Wermelinger, A. Bidiville, R. Spolenak, and J. Michler, *In situ compression tests on micron-sized silicon pillars by Raman microscopy—Stress*

measurements and deformation analysis. Journal of Materials Research, 2008. **23**(11): p. 3040-3047.

19. Z. Iqbal and S. Veprek, *Raman-scattering from hydrogenated micorcrystalline and amorphous-silicon*. Journal of Physics C-Solid State Physics, 1982. **15**(2): p. 377-392.

20. Y. Gogotsi, C. Baek, and F. Kirscht, *Raman microspectroscopy study of processing-induced phase transformations and residual stress in silicon*. Semiconductor Science and Technology, 1999. **14**(10): p. 936-944.

21. H. Tang and I.P. Herman, *Raman microprobe scattering of solid silicon and germanium at the melting temperature*. Physical Review B, 1991. **43**(3): p. 2299-2304.

22. T. Wermelinger, C. Borgia, C. Solenthaler, and R. Spolenak, *3-D Raman spectroscopy measurements of the symmetry of residual stress fields in plastically deformed sapphire crystals*. Acta Materialia, 2007. **55**(14): p. 4657-4665.

23. D.S.e.L. SL, *Lince v. 231*. 1998, FB Materials Science, Ceramics Group: TU Darmstadt.

24. L. Mariucci, A. Pecora, G. Fortunato, C. Spinella, and C. Bongiorno, *Crystallization mechanisms in laser irradiated thin amorphous silicon films*. Thin Solid Films, 2003. **427**(1-2): p. 91-95.

25. J.S. Im, H.J. Kim, and M.O. Thompson, *Phase-transformation mechanisms involved in excimer-laser crystallization of amorphous-silicon films.*. Applied Physics Letters, 1993. **63**(14): p. 1969-1971.

26. H. Watanabe, H. Miki, S. Sugai, K. Kawasaki, and T. Kioka, *Crystallization process of polycrystalline silicon by KrF excimer-laser annealing.*. Japanese Journal of Applied Physics Part 1-Regular Papers Short Notes & Review Papers, 1994. **33**(8): p. 4491-4498.

27. T. Sameshima, K. Saitoh, M. Sato, A. Tajima, and N. Takashima, *Crystalline properties of laser crystallized silicon films*. Japanese Journal of Applied Physics Part 2-Letters, 1997. **36**(10B): p. L1360-L1363.

28. M. Miyasaka and J. Stoemenos, *Excimer laser annealing of amorphous and solid-phase-crystallized silicon films*. Journal of Applied Physics, 1999. **86**(10): p. 5556-5565.
29. D.H. Choi, E. Sadayuki, O. Sugiura, and M. Matsumura, *Lateral growth of poly-Si by excimer laser and its thin-film-transistors application*. Japanese Journal of Applied Physics Part 1-Regular Papers Short Notes & Review Papers, 1994. **33**(1A): p. 70-74.
30. D.H. Choi, K. Shimizu, O. Sugiura, and M. Matsumura, *Drastic enlargement of grain-size of excimer-laser-crystallized polysilicon films*. Japanese Journal of Applied Physics Part 1-Regular Papers Short Notes & Review Papers, 1992. **31**(12B): p. 4545-4549.
31. H.R. Philipp and E.A. Taft, *Optical constants of silicon in the region 1 to 1 eV*. Physical Review, 1960. **120**(1): p. 37-38.
32. J. Menendez and M. Cardona, *Temperature-dependence of the 1st-order Raman scattering by phonons in Si, Ge and A-SiN - anharmonc effects*. Physical Review B, 1984. **29**(4): p. 2051-2059.
33. R. Tsu and J.G. Hernandez, *Temperature-depenence of silicon Raman lines*. Applied Physics Letters, 1982. **41**(11): p. 1016-1018.
34. Anastass.E, A. Pinczuk, E. Burstein, F.H. Pollak, and M. Cardona, *Effect of static uniaxial stress on Raman spectrum of silicon*. Solid State Communications, 1970. **8**(2): p. 133-&.
35. K. Brendel, N.H. Nickel, P. Lengsfeld, A. Schopke, I. Sieber, M. Nerding, H.P. Strunk, and W. Fuhs, *Excimer laser crystallization of amorphous silicon on metal coated glass substrates*. Thin Solid Films, 2003. **427**(1-2): p. 86-90.
36. M. Nerding, R. Dassow, S. Christiansen, J.R. Kohler, J. Krinke, J.H. Werner, and H.P. Strunk, *Microstructure of laser-crystallized silicon thin films on glass substrate*. Journal of Applied Physics, 2002. **91**(7): p. 4125-4130.
37. S. Christiansen, P. Lengsfeld, J. Krinke, M. Nerding, N.H. Nickel, and H.P. Strunk, *Nature of grain boundaries in laser crystallized polycrystalline silicon thin films*. Journal of Applied Physics, 2001. **89**(10): p. 5348-5354.

38. S. Loreti, M. Vittori, L. Mariucci, and G. Fortunato, *Characterisation of excimer laser crystallised polysilicon by X-ray diffraction and by channeling contrast in a scanning electron microscope.* Solid State Phenomena, 1999. **67-8**: p. 181-186.
39. V.L. Borblik, *Determination of 'bisotropic' stresses in layered semiconductor structures from Raman light scattering data.* Journal of Physics-Condensed Matter, 2007. **19**.
40. M. Chandrasekhar, J.B. Renucci, and M. Cardona, *Effects of interband excitation on Raman phonons in heavily doped N-Si.* Physical Review B, 1978. **17**(4): p. 1623-1633.
41. A. Ballantyne, H. Hyman, C.L. Dym, and R. Southworth, *Response of lithographic mask structures of intense repetitively pulsed X-rays - thermal-stress analysis.* Journal of Applied Physics, 1985. **58**(12): p. 4717-4725.
42. M. Mondol, H.Y. Li, G. Owen, and H.I. Smith, *Uniform-stress tungsten on X-ray mask membranes via He-backside temperature homogenization.* Journal of Vacuum Science & Technology B, 1994. **12**(6): p. 4024-4027.
43. C.S. Montross, H. Yokokawa, and M. Dokiya, *Thermal stresses in planar solid oxide fuel cells due to thermal expansion differences.* British Ceramic Transactions, 2002. **101**(3): p. 85-93.
44. Y. Guyot, C. Malhaire, M. LeBerre, B. Champagnon, A. Sibai, E. Bustarret, and D. Barbier, *Micro-Raman study of thermoelastic stress distribution in oxidized silicon membranes and correlation with finite element modeling.* Materials Science and Engineering B-Solid State Materials for Advanced Technology, 1997. **46**(1-3): p. 24-28.
45. J.H. Chae, J.Y. Lee, and S.W. Kang, *Measurement of thermal expansion coefficient of poly-Si using microgauge sensors.* Sensors and Actuators a-Physical, 1999. **75**(3): p. 222-229.
46. F.C. Nix and D. MacNair, *The thermal expansion of pure metals copper, gold, aluminum, nickel, and iron.* Physical Review, 1941. **60**(8): p. 597-605.
47. A.K. Sinha, H.J. Levinstein, and T.E. Smith, *Thermal-stresses and cracking resistance of dielectric films (SiN, Si3N4 and SiO2) on Si substrates.* Journal of Applied Physics, 1978. **49**(4): p. 2423-2426.

48. M.A. Elkhakani and M. Chaker, *Physical-properties of the X-ray membrane materials.* Journal of Vacuum Science & Technology B, 1993. **11**(6): p. 2930-2937.
49. S.J. Hwang, Y.D. Lee, Y.B. Park, J.H. Lee, C.O. Jeong, and Y.C. Joo, *In situ study of stress relaxation mechanisms of pure Al thin films during isothermal annealing.* Scripta Materialia, 2006. **54**(11): p. 1841-1846.
50. D.S. Gardner and P.A. Flinn, *Mechanical-stress as a function of temperature for aluminum-alloy films.* Journal of Applied Physics, 1990. **67**(4): p. 1831-1844.

4 Correlating Raman Peak Shifts with Phase Transformation and Defect Densities: a Comprehensive TEM and Raman Study on Silicon

So far, the chapter showed that Raman microscopy is an accurate tool to measure stresses in one and two dimensional cases. In these rather simple cases the stress tensor is clearly defined and it is possible to correlate Raman peak shifts with occurring stresses. This chapter concentrates on microstructural changes which can be mapped by Raman microscopy. In this particular experiment a focused ion beam (FIB) was used to extract a silicon lamella from the center of an indent. A part of the Ga^+ ions which were used for cutting the lamella are implanted into the silicon lamella and alter the silicon Raman signal. Therefore, the chapter deals also with the question how the sample preparation directly influences the results. The here presented work is accepted as a full length article: Wermelinger T., Spolenak R.; *Correlating Raman peak shifts with phase transformation and defect densities: a comprehensive TEM and Raman study on silicon*, Journal of Raman spectroscopy 40 (2009)

4.1 Abstract

Silicon is the most often used material in micro electro mechanical systems (MEMS). Detailed understanding of its mechanical properties as well as the microstructure is crucial for the reliability of MEMS devices. In this paper, we investigate the microstructure changes upon indentation of single crystalline (100) oriented silicon by transmission electron microscopy (TEM) and Raman microscopy. TEM cross sections were prepared by focused ion beam (FIB) at the location of the indent. Raman microscopy and TEM revealed the occurrence of phase transformations and residual stresses upon deformation. Raman microscopy was also used directly on the cross-sectional TEM lamella and thus microstructural details could be correlated to peak shape and peak position. The results show, however, that due to the implanted Ga^+ ions in the lamella the silicon Raman peak is shifted significantly to lower wavenumbers. This hinders a quantitative analysis of residual stresses in the lamella. Furthermore, Raman microscopy also possesses the ability to map deformation structures with a lateral resolution in the submicron range.

4.2 Introduction

The mechanical and electrical properties of silicon (Si) are of high interest as it is the most industrially important semiconducting material. Due to the development of Micro Electro-Mechanical Systems (MEMS) the mechanical properties as well as the microstructure of silicon became a focus of research. As the dimensions of the devices are in the micro- and submicrometer range, new methods for testing the properties of silicon at this scale had to be established.

One of these methods is micro- and nanoindentation which is often used to investigate the mechanical deformation behavior as well as occurrence of pressure-induced phase transformation in silicon [1-5].Another approach for testing the mechanical properties in small dimensions was presented by B. Moser *et al.*[6] They performed micro-compression tests on pillars with a diameter ranging from 1 to 50 μm. Combining micro-compression tests on silicon pillars with Raman microscopy allowed the detection of cracks prior to failure[7]. Compression tests of silicon nanoparticles in TEM enabled to observe elastic and plastic deformation in situ [8]. Indentation is a very accurate method to obtain several characteristic material properties such as Young's modulus and hardness. Mechanical properties, however, need to be correlated to microstructure to allow an interpretation of deformation mechanisms. In this work the interpretation will be accomplished by Raman microscopy and TEM analysis of cross sections as prepared by FIB.

Over the last 15 years focused ion beam (FIB) has become one of the most effective methods used in preparation of transmission electron microscopy (TEM) samples. This method largely increased the possibilities for detailed examinations of the microstructure of materials as it enables the visualization of the microstructure. One great advantage is its high positioning accuracy of the ion beam for the preparation of cross-sectional samples. It is well known that the preparation of such cross-sectional samples leads to implantation of ions into the sample surface. The implanted ions cause severe damage in the surface of the sample down to a depth of 10-30 nm [9-14]. Tamura *et al.* found that above a critical dose of implanted Ga^+ ions by means of FIB in silicon (of about 8 to $10*10^{13}$ ions/cm^2) an amorphous layer silicon is formed at the surface [15]. Bourgoin *et al.*[16] could show the amorphous layer formation in silicon by ion implantation for the case of argon ions. Recently, FIB cut micropillars have become a popular model system for size effects in mechanical properties [17-20]. There are still discussions about the influence of the implanted ions on the sample itself as it could be the cause for residual stresses. This may alter the features observed by TEM or also possibly alter the deformation behavior of nanostructures, which have been prepared by FIB. Bei *et al.* [21] found that implanted Ga^+ ions due to FIB-milling can introduces damages which are needed for dislocation nucleation during deformation.

Raman microscopy has proven to be an accurate method for analyzing certain microstructural aspects such as grain size, phase transformations and residual stresses in silicon [1, 22-24]. The big advantage of Raman microscopy is the possibility of measuring and mapping elastic strain fields as well as microstructural changes with a lateral resolution in the submicrometer range. Recent developments enable the determination of the stress tensor [25, 26]. Moreover, confocal Raman spectroscopy can be performed to map 3 dimensional stresses in transparent materials such as sapphire [27]. As such it is a non-destructive complementary technique to TEM.

The combination of FIB as a method for sample preparation and Raman microscopy for mapping has the potential to give new insights in the microstructure of deformed material. In this work phase transformations and residual stresses around a micro indent were studied. A cross-sectional lamella from the center of the indent was studied by TEM as well as Raman microscopy.

4.3 Experimental

A (001) oriented silicon wafer was plastically deformed by means of Vickers microindention. The maximal applied force was 0.2 N with a loading and unloading rate of 0.05 N/s. The radius of the indenter varied from 0.16 μm to 1.1 μm depending on the direction. The resulting indent with an approximate diameter of 2.5 μm and a depth of 0.26 μm (see Figure 4.1) was scanned by means of atomic force microscope (AFM) (CRAFM 200, WITec GmbH, Germany). The line indicates the position where a lamella was extracted by FIB for further examinations with a TEM. The circle marks a pile-up at the border of the indent.

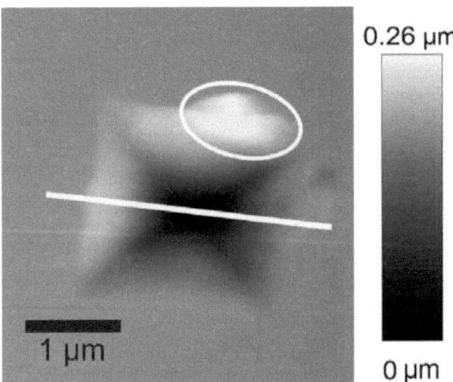

Figure 4.1 AFM image of the Vickers indent placed on the (001) oriented silicon wafer. The indent had a diameter of around 2.5 μm. The line indicates the position of the TEM-lamella extracted from FIB. The circle marks a pile-up of silicon at the border of the indent.

A confocal Raman microscope (CRM 200, WITec GmbH, Germany) was used to map the indent and the surrounding area to detect occurring residual stresses and phase transformations. The microscope was equipped with a blue laser with a wavelength λ of 442 nm. The measurement was performed in a backscattering mode with a 100x objective which had a numerical aperture (NA) of 0.9. The resulting lateral resolution of the system was 300 nm.

As silicon being a highly Raman active material it was possible to perform Raman mappings on the TEM-lamella which was cut out from the indent. The lamella was 12 x 13 μm^2 wide and about 200 nm thick.

Seven 10 x 10 μm^2 big areas and one 5 x 5 μm^2 of a (001) oriented Si wafer were irradiated with Ga^+ ions by means of FIB to analyze the influence of the implanted ions on the silicon Raman peak. For five of the 10 x 10 μm^2 big areas the energy was kept constant at 30 keV while the current was set to the following different values: 10, 40, 80, 150 and 300 pA. The one 5 x 5 μm^2 big area was also irradiated with 30 keV but with a current of 1 pA. This led to ion fluences between $1.3*10^{14}$ and $9.4*10^{15}$ Ga^+/cm^2. Two further areas were irradiated with 80 pA with a energy of 2 keV and 5 keV, respectively. In all cases the irradiation duration was 5 seconds. The irradiated areas were analyzed by Raman microscopy with different laser powers of 7.6 mW, 14 mW and 20 mW.

Further, a (001) oriented silicon wafer was irradiated with helium ions to produce defects in the silicon bulk material. The irradiation was performed with a fluence of $5 * 10^{16}$ $ions/cm^2$ and an energy of 2 MeV. The irradiation direction was inclined by 8° to the surface normal to avoid channeling which would strongly increase the penetration depth of the ions [28]. After the irradiation, the silicon wafer was cracked through the center of the irradiated area to be scanned linearly along the (001) direction of the crack (see Figure 4.2) by the confocal Raman microscope.

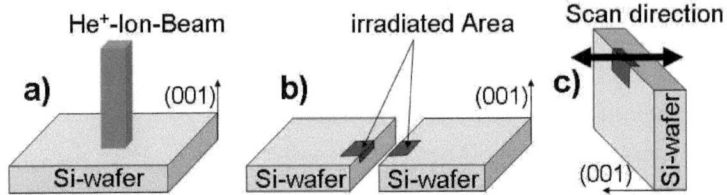

Figure 4.2 a) Irradiation of a silicon wafer with He⁺ ions with a fluence of $5*10^{16}$ ions/cm² and energy of 2 MeV. b) Cracking of the wafer through the irradiated area. c) Line scans parallel to the (001) direction on the cracked surface.

The program SRIM (Stopping and Range of Ions in Matter) allows simulating the behavior of ions [29] such as collisions events as well as the possible penetration depth of the ions and was utilized to simulate the interaction of helium ions with silicon.

4.4 Results

Figure 4.3 a) shows the result of the Raman map around the Vickers indent. From the Raman map four different areas in the indent and its vicinity can be distinguished. The labels "a" to "d" in Figure 4.3 a) mark the regions where a characteristic spectrum of each area was extracted which can be seen in Figure 4.3 b). Spectrum "a" belongs to unstressed pristine Si-I peak and can be found 3-4 µm away from the indent. The peak position of the unstressed pristine silicon is at 519.6 to 519.8 cm⁻¹. Around the indent the Si-peak is shifted to higher wavenumbers (see spectrum "b") which refers to an area of compressive stresses. Spectrum "c" was taken from the pile-up at the border of the indent (see Figure 4.1). The Si-peak appears at 507 cm⁻¹ which indicates a phase transformation from the cubic Si-I to the hexagonal Si-IV [2, 22]. Direct phase transformation from Si-I to Si-IV is known to require a high level of shear deformation[1, 30]. This phase only occurred at the pile-up at the border of the indent, therefore, in an area where the material was exposed to high shear deformation. In the center of the indent (see spectrum "d") the Raman signal shows amorphous silicon which has a Raman peak position at around 475 cm^{-1}. In comparison with the Raman signal of the crystalline silicon (spectrum "a") the amorphous silicon has a very high peak width. The results agree with the findings of Kailer *et al.* [1] that directly below the indent a fast load release leads to a phase transformation from Si-I to amorphous silicon.

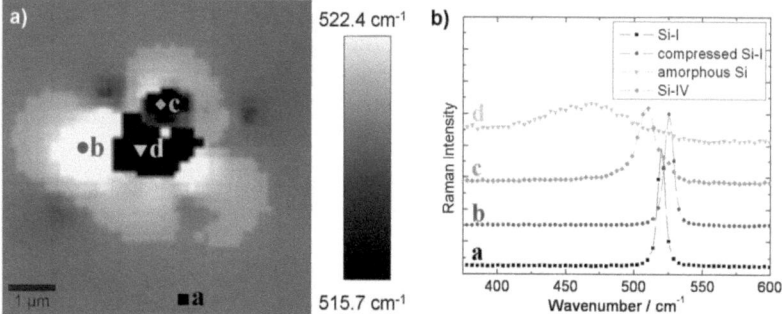

Figure 4.3 a) Mapping of the shift of the silicon Raman peak around the indent. 4 different zones, marked by labels, are visible. b) Characteristic Raman spectra of each zone; a) pristine Si-I, b) spectrum of Si-I in a compressive state, c) Si-IV belonging to the region of the pile-up, d) Amorphous silicon belonging to the center of the indent.

A FIB lamella from the center of the indent (see Figure 1) was extracted to visualize and analyze the deformed microstructure below the indent by means of TEM. Figure 4.4 shows a TEM image of the indent. The dark band "a" along the sample is a platinum layer which is needed for cutting out the lamella. Arrow "b" points to the center of the indent. Just below the indent, marked by arrow "c", is a zone which shows no contrast. This zone has a maximal diameter of around 0.6 μm. The inlay in Figure 4.4 shows a diffraction pattern taken from area "c". The diffraction pattern shows only wide diffused rings which correspond to amorphous silicon. Therefore, the TEM image is consistent with the Raman mapping. The amorphous area, surrounded by a heavily disturbed zone roughly 1 μm in diameter as shown by arrow "d", has a high density of dislocations. One can assume that in this zone the compressive stress was not high enough to convert it to amorphous silicon.

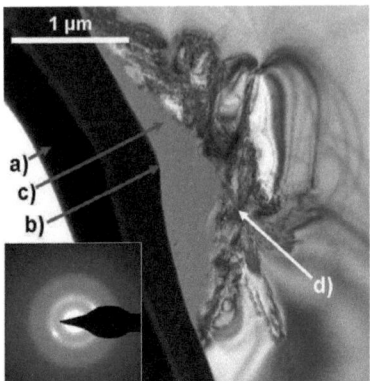

Figure 4.4 TEM image of the FIB lamella extracted from the center of the indent. a) platinum layer. b) center of the indent. c) area of amorphous silicon. d) area with high density of dislocation. The inlay shows a diffraction pattern from the amorphous area (area c)) beneath the indent.

The high Raman activity of silicon allows mapping of very thin films and layers such as a FIB lamella. Figure 4.5 shows the Raman maps around the indent. The arrows in both figures mark the center of the indent. The black region on the left side of the mappings denotes to the border of the lamella. Figure 4.5 a) shows the shift of the peak position of the silicon peak. From the scale bar of the mapping, which ranges between 499 cm^{-1} and 515 cm^{-1}, it is visible that the Raman peak is strongly shifted to lower wavenumbers in comparison with the pristine silicon peak which is at 520 cm^{-1}. The highest peak shifts were found at the center of the indent. The area of these large shifts has a diameter of 0.7 μm and is related to the occurrence of the amorphous silicon phase. The surrounding region of the indent has a uniform peak shift to lower wavenumbers in comparison to the normal silicon. Figure 4.5 b) shows the peak width (full width at half maximum) of the silicon peak. In the vicinity of the indent the peak width is significantly larger than in the regions further away, where the peak width is homogeneous. The area with the broad peaks (see Figure 4.3 b)) is significantly larger than the area with the amorphous silicon.

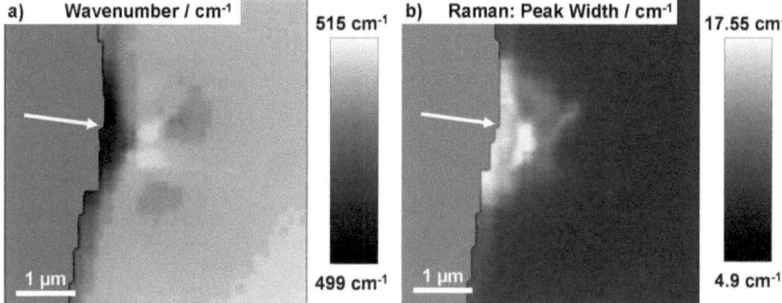

Figure 4.5 Raman map of the FIB lamella around the indent (see Figure 4.4). The arrow marks the center of the indent. a) Map of the shift of the silicon Raman peak. b) Map of the peak width of the silicon Raman peak.

FIB equipped with a Ga^+ ion beam, was used to cut out the lamella. To simulate the damage caused by ion implantation within the lamella during the sample preparation a silicon wafer was irradiated with different doses of Ga^+ ions. Figure 4.6 a) shows spectra measured from the areas irradiated with an energy of 30 keV. With increasing fluence of ions the crystalline silicon peak decreases while an amorphous silicon peak appears. At the highest fluence of $9.4*10^{15}$ ions/cm^2 the crystalline peak is almost completely vanished. The inlay shows the crystalline to amorphous peak-ratio as a function of the irradiation dose. Above a fluence of $3.1*10^{14}$ ions/cm^2 the amorphous peak is dominating the Raman spectrum. In Figure 4.6 b) the spectra of the areas with a fluence of $2.5*10^{15}$ ions/cm^2 irradiated with 2, 5 and 30 keV are visible. Only the sample irradiated with 30 keV has an observable amorphous Raman peak. The two other samples seem to be totally crystalline.

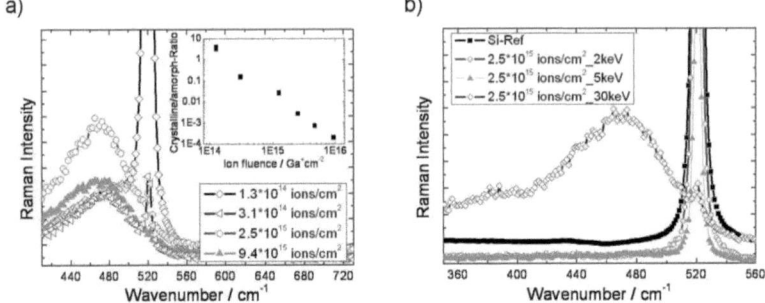

Figure 4.6 a) silicon Raman spectra for different fluences irradiated with an energy of 30 keV. The inlay shows the ratio between the crystalline peak and the amorphous peak. At the highest irradiation dose almost only amorphous silicon is detected. b) shows the same fluence of $2.5*10^{15}$ Ga^+ ions/cm^2 but varying energy of 2 keV, 5 keV and 30 keV. Only the sample irradiated with an energy of 30 keV has an amorphous silicon Raman peak.

In the areas with the lowest fluence of $1.3*10^{14}$ Ga^+ ions/cm^2 the Raman spectrum was measured as a function of time and Raman laser power. Figure 4.7 shows the result of the measurements. With increasing laser power the Raman peaks of the reference measurements on the non-irradiated spot as well as on the irradiated measurements are shifted to lower wavenumbers. The shift between the reference measurement performed with 20 mW and the reference measurement performed with 7.6 mW is in the order of 0.5 cm^{-1}. At the beginning of the measurements of the irradiated samples the peak position was not stable. Moreover, they did not show a consistent behavior. After 300 seconds the peak positions of the irradiated samples seemed to become stable. The measurements on the irradiated spots were always below to their corresponding reference measurement. The shift between the irradiated sample and the corresponding reference measurement was in the stable region (above 300 seconds) in the order of 0.3-0.45 cm^{-1}.

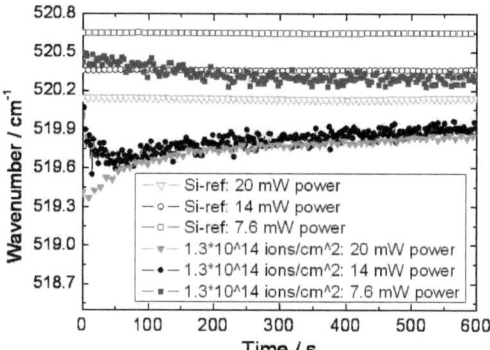

Figure 4.7 Time scans measured on the area irradiated with $1.3*10^{14}$ Ga$^+$ ions/cm^2 and on non-irradiated, reference spots. The measurements were performed at 7.6 mW, 14 mW and 20 mW power of the Raman laser.

Another approach to see the influence of irradiated ions on the Raman spectrum was to implant He$^+$ ions in to a silicon wafer. Raman line scans with a length of 30 μm were performed on the cracked surface of the wafer in an area where the wafer was irradiated with the He$^+$ ions (see Figure 4.2). Figure 4.8 shows the peak position as a function of the depth, starting form the edge of the surface.

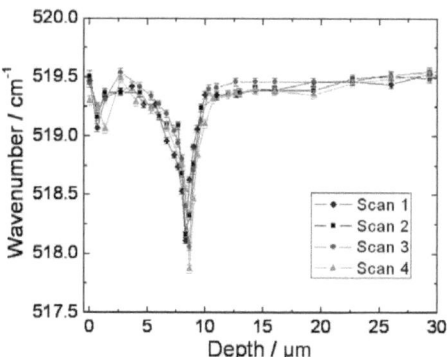

Figure 4.8 Line scan along the cross-section of the irradiated area of the silicon wafer (see Figure 4.2). All performed scans show a maximal peak shift at a depth of 8 μm.

All four scans exhibit a consistent behavior. On the surface the peak position is at 519.5 cm^{-1} and decreases by 0.5 cm^{-1} to a depth of about 1 μm. With increasing depth the peak position increases before decreasing again. At a depth of around 8 μm the line scans show the maximal peak shift to 518 cm^{-1}. Between 8-10 μm the peak position increases strongly to around 519.2 cm^{-1}. In the region between 10 and 30 μm the peak position is slowly increasing to 519.5 cm^{-1}.

4.5 Discussion

The main focus of this work is to correlate microstructural details as observed by TEM to changes in the Raman spectrum. The occurrence of an amorphous phase can thus be directly correlated to a strong peak shift and peak broadening. A high defect structure, on the other hand, can be correlated to the observation of peak broadening alone. On the other, the absolute value of peak shifts on a TEM lamella cut by FIB is more complex as both elastic strain as well as a change in defect density as caused by ion implantation can contribute to a Raman peak shift.

Figure 4.5 a) shows that a general peak shift to lower wavenumbers can be observed. Such a peak shift could be indicated by residual tensile stresses in the lamella. The implantation of Si$^+$ ions into a silicon wafer with an energy of 1.5 MeV and a doses ranging from $1*10^{11}$ to $1*10^{15}$ Si$^+$ ions/cm^2 lead to compressive residual stresses up to 120 MPa in the surface layer [31]. If the shift were solely caused by internal stresses it would correspond to tensile stresses ranging from 2.5 to 10 GPa [32]. Tensile stresses in this range would most likely lead to fracture of the lamella. Residual stresses are thus considered to be of secondary importance with regards to Raman peak shifts. Figure 4.5 b) presents the increased peak width of the Raman signal around the indent. One part of the area with the increased peak width can be explained with the presence of amorphous silicon as it exhibits a broad peak (see Figure 4.3 b)). The area of increased peak width in the membrane is larger than the area of the amorphous silicon. Therefore, also other effects than the phase transformation to amorphous silicon, lead to an increase in peak width of the Raman signal.

Figure 4.9 shows the superposition of the TEM image of the FIB lamella (see Figure 4.4) and the Raman map of the peak width around the indent (see Figure 4.5 b). To increase the visibility, the level of transparency of the TEM image in the sequence of images changes from high to low. The sequence of images clearly shows that the distribution of the dislocations fits very well with the Raman peak width. From literature it is known that disorder of silicon leads to a change of peak position as well as peak shape [33, 34]. The area with a high density of dislocations beneath the indent leads to a peak broadening while the peak position is not changed significantly. One can conclude that analyzing the peak width makes it possible to qualitatively map the dislocation density in silicon while the peak shift is superimposed by other effects.

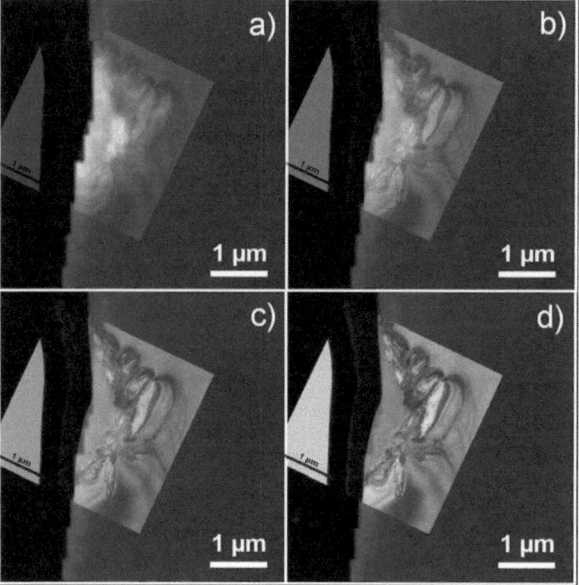

Figure 4.9 Superposition of the map of the peak width (see Figure 4.5 b)) and the TEM image of the indent (see Figure 4.4). For better visibility, in the sequence of images the transparency level of the TEM image ranges form high transparency (a)) to low transparency (d)).

Another explanation for a systematic peak shift could be caused by laser heating of the lamella during the Raman measurement, because of the high power density of the laser spot. The Raman peak position is temperature dependant and heating up a sample leads to a shift of the signal to lower wavenumbers [35]. The results of the measurements with different laser power (see Figure 4.7) indicates that the silicon peak is shifted to lower wavenumber with increasing laser power. This is due to a heating of the sample with the laser. This is clear evidence that sample heating can be an issue for such kind of experiments. Figure 4.10 a) shows the shift of the silicon peak of the whole FIB lamella. The shift of the silicon peak is, beside the region of the indent, marked by the white circle, and a small strip on the right side of the lamella, fairly uniform and ranges between 512 cm^{-1} and 514 cm^{-1}. Figure 4.10 b) presents the intensity of the silicon peak for the same measurement. The intensity mapping of the lamella exhibits significant variations with certain areas exhibiting signal intensity 10 times higher than in other areas. One reason for this behavior is the non uniform thickness of the lamella. The lamella has a "frame" which is thicker than the inner part. This frame is originating from the preparation process and is 1 to 2 μm wide. In Figure 4.10 b) the outlines of the frame are visible due to the increased signal from this region. Another reason for differences in the signal intensity is the problem of focusing on the lamella. The better the laser is focused on the sample the higher is the measured signal intensity. The laser is not always perfectly focused on the sample during the scan because the sample is tilted due to the attachment to a sample holder. This leads to differences in the signal intensity within the lamella. This behavior can be observed in the inner part of the lamella (see Figure 4.10 b)) where on the upper side the intensity signal is low in comparison with the signal in the lower left corner. High signal intensity corresponds also to a high energy density which is brought into the sample. If the signal intensity changes drastically, as it does in this case, while the peak position of the signal is not shifted, one can conclude that even a high laser energy density does not lead to local shifts induced by local heating effects. Nevertheless, one has to point out that a general small peak shift to lower wavenumbers due to a homogeneous heating of the sample cannot be entirely excluded.

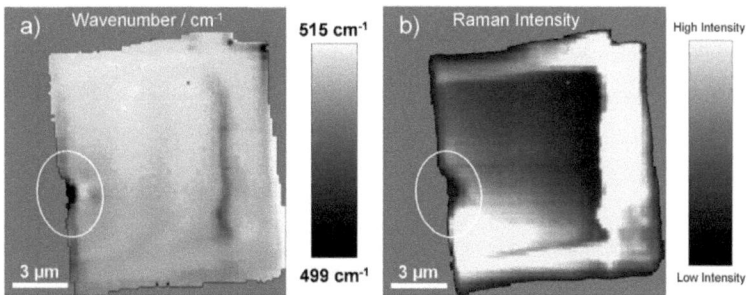

Figure 4.10 a) mapping of the whole FIB lamella of the Raman silicon peak position. b) mapping of the intensity of the Raman silicon peak. The white circle denotes the region of the indent (see Figure 4.4).

Similar to the findings in literature[15], the irradiation of Ga$^+$ ions leads to an amorphization of the surface. From the experiments performed with an energy of 30 keV (see Figure 4.6 a)) one can conclude that the fluence for a total transformation from crystalline to amorphous silicon is in the range of $9.4*10^{15}$ Ga$^+$ ions/cm^2. But already at a much smaller fluence of $1.3*10^{14}$ Ga$^+$ ions/cm^2 a partial amorphization of the material is detectable. This is in contrast to performed experiments with 2 and 5 keV (see Figure 4.6 a)). At an energy of 30 keV and a fluence of $2.5*10^{15}$ Ga$^+$ ions/cm^2 a large amorphous peak is visible. Whereas the same fluence in the case of the 2- and 5 keV-experiments does not lead to any significant amorphization of the material.

Richter, Wang and Ley described the Raman spectrum of microcrystalline silicon films [34]. They showed that the peak position as well as the peak shape depends on the crystallite size. Below a certain grain size the peak is shifted to lower wavenumbers. This behavior can be explained by the phonon confinement model (PCM). It assumes that phonons are not infinite as in the single crystalline case but restricted by grain boundaries. The restriction causes a relaxation of the phonon wave vector. The smaller the crystallite size the higher is the peak shift. According to the literature a crystallite size of 3 nm corresponds to a Raman peak at 512 cm^{-1} [24]. Ion implantation leads to disturbed cluster as well as undamaged crystalline regions. Similar to the case of microcrystalline silicon below a certain size of undamaged crystals the Raman peak is shifted to lower wavenumbers. Figure 4.7 shows such a behavior. In the stable region above 300 seconds the peak position of the irradiated samples is shifted 0.3 to 0.45 cm^{-1} to lower wavenumbers regardless of the used laser power. The behavior of the irradiated samples at the beginning of the time scans is not fully understood and needs further examinations.

To verify that the implanted ions do lead to a shift of the Raman peak, another model system was looked for. The aim was to have a system where peak shifts were detectable but did not show strong amorphization of the substrate material. He$^+$ ions as well as the high irradiation energy of 2 MeV was chosen to create defects not at the surface but within the bulk material. Although it is known that high fluences lead to helium bubbles in the substrate[36]. To detect the influence of the implanted He$^+$ ions Raman depth scans were performed. These kinds of measurements are not possible to perform with Ga$^+$ ions. Figure 4.8 shows the Raman peak shift as a function of the depth in a He$^+$ ion irradiated area. To prove that the shift is caused by ion bombardment a SRIM calculation was performed (see Figure 4.11). The irradiation creates two different kinds of disorder in the silicon bulk material. The penetrating ions cause defects in the crystal due to collisions events with the atoms of the crystal. Figure 4.11 a) shows the number of occurring collisions events as a function of the penetration depth. The number of collisions directly correlates to the number of defects. The figure shows that the maximum of defects is at a depth of around 7 μm. The other kind of disorder is the stopping of the penetrating ions (see Figure 4.11 b). The SRIM simulation predicts that great majority of atoms stop in a depth between 6 and 8 μm. The peak shift of the Raman line scans correspond to the results of the SRIM simulation. The maximum of defects and implanted ions, where the undamaged crystalline regions are the smallest, are at the same depth as the maximum of the peak shift in the Raman scans. This clearly shows that implanted ions effect the peak position of the Raman signal.

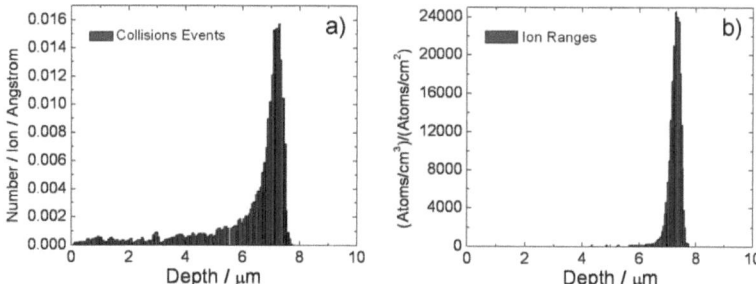

Figure 4.11 SRIM simulation[29] for the irradiation of He$^+$ ions with an energy of 2 MeV into silicon. a) numbers of defects caused by the irradiation with a maximum at a depth of around 7 μm. b) stopping range of the He$^+$ ions in the silicon. Most of the ions stop between 6-8 μm within the silicon.

According to Rubanov and Munroe the implanted dose of Ga$^+$ ions, using a beam voltage of 30 keV and a beam current of 150 pA (see table 1 and equation 3) [9], is $4.2*10^{16}$ ions/cm^2. SRIM simulation showed that the penetration depth of the Ga$^+$ ions is not exceeding 30 nm and causes ~400 vacancies/ion, using an incident angle of 85°. In the case of the implanted He$^+$ ions a fluence of $5*10^{16}$ ions/cm^2 was used. According to SRIM simulation every He$^+$ ion caused 200 vacancies. The fluence as well as the vacancies/ions in both cases of ion implantation is comparable. This shows that the implanted Ga$^+$ ions into the lamella has to affect the Raman signal. The stopping range of the Ga$^+$ ions is between 0 nm and 30 nm whereas the stopping range of the He$^+$ ions is between 6 and 8 μm. As a consequence the defect density per volume is in the case of the Ga$^+$ ions higher. This could be the reason for the higher Raman peak shifts of the TEM lamella in comparison with the measurements performed on the He$^+$ ion implanted sample.

The SRIM simulation and the Raman line scans clearly showed the influence of implanted ions on the Raman signal. During the preparation process by means of FIB Ga$^+$ ions are implanted into the surface of the lamella. The vertical dark strip which indicates a higher peak shift in Figure 4.10 a) corresponds to an area where the dose of implanted ions is known to be higher due to sample preparation. Residual stresses as well as heating effects can be neglected as the main reason for the general peak shift to lower wavenumbers. Therefore, the main portion of the peak shift in the FIB lamella is induced by the implanted Ga^+ ions.

4.6 Conclusions

A Vickers indent was placed normal to the (001) plane of silicon. Residual stresses and phase transformations were studied around the indent by means of Raman microscopy. In the center of the indent amorphous silicon was found while a pile-up at the border of the indent could be assigned to a phase transformation from the cubic Si-I to hexagonal Si-IV. The surrounding area of the indent is dominated by compressive stresses.

A cross-sectional lamella from the indent was extracted by FIB. The deformed microstructure was analyzed with TEM. As silicon is very Raman active it was possible to do Raman mapping on the lamella. One result of the mapping is that the silicon peak is generally shifted strongly to lower wavenumbers. Internal stresses could lead to shifts but the contribution only plays a minor role in this case. Heating leads to a uniform shift to lower wavenumbers but no local heating effects could be observed. Ga^+ ions were implanted with different fluences and energies. In combination with He^+ ion implantation in silicon and SRIM simulations they showed that defects and implanted ions effect the Raman peak position of silicon. It is known that during the sample preparation of the lamella Ga^+ ions are implanted into the surface. These implanted ions are causing a peak shift. Moreover, areas in the lamella which are supposed to have a higher concentration of implanted ions showed larger peak shifts.

Due to the large peak shift induced by the implanted ions it was not possible to analyze residual stresses in the lamella. Nevertheless, the peak width distribution was in very good agreement with the dislocation distribution near the indent. Our results show that Raman microscopy is able to qualitatively map the distribution of the dislocation density.

4.7 Acknowledgement

The authors thank C. Solenthaler for the TEM analysis, S. Olliges for the SRIM simulation as well as T. Frey for his work on the implantation of Si^{2+}-ions in silicon. The author acknowledges the work of R. Nyilas who performed the microdiffraction experiments and also analyzed the data. This work was supported by the ETH Research Grant TH -39/05-1.

4.8 Appendix

4.8.1 Synchrotron X-Ray Laue Microdiffraction on He^+-ion Implanted Silicon

Beside the Raman measurements also synchrotron X-ray Laue microdiffraction experiments were performed. These experiments allowed analyzing the mechanical stresses induced by the implanted ions. A big advantage of the method is that it is able to measure in ideal cases all six components of the strain tensor with a lateral resolution in the micrometer range. Direct measurement of mechanical stresses by means of Raman microscopy was not possible in this case since the implanted ions not only cause stresses but also lead to a phonon confinement. Until now, it was not possible to separate the two effects from each other.

The microdiffraction experiments were conducted at the MicroXAS (X05LA) beam line at the Swiss Light Source (SLS) providing micron-focused X-rays from an undulator insertion device (photon beam energy of 2.4 GeV at a beam current of 400 mA) in the energy range between 5-18 keV. The pre-optics and KB mirrors [37] were optimized to achieve a beam spot size of 1.1x1.4 micron (horizontal x vertical). The Laue diffraction patterns were collected using the Pilatus 6M photon counting area detector (developed at the SLS) which is based on the CMOS hybrid pixel technology operating in single–photon counting mode [38, 39]. X-rays are directly transformed into electric charge in the silicon sensor. The signals are amplified and processed in the CMOS readout chips. This technology allows collecting noise free data with a maximum dynamic range of 20 bit at speeds ranging from 10-300 Hz. In comparison, area detectors based on charge coupled devices (CCDs) or imaging plates have a restricted dynamic range of <17 bit with a relatively high intrinsic background and long readout times. Pilatus detectors feature an adjustable lower energy threshold which allows for energy discrimination which is important for experiments where most of the desired information is within weak diffraction intensities.

Using a fixed diffraction geometry ($2\theta=90°$) the sample was translated relative to the incoming beam to collect Laue diffraction frames at different sample positions to scan over the He-Ion implanted region covering a depth of 8-10μm. A total of 441 Laue diffraction frames were collected corresponding to a scanned area of 20x20 μm through the depth of the He-Ion implanted region. The Laue diffraction patterns have been indexed using the software package XMAS [40]. The software allows to index overlapping Laue patterns originating from different grains within the illuminated diffraction volume.

Figure 4.12 shows the indexed Laue diffraction frame originating from the silicon sample at room temperature. The absolute peak positions show the crystallographic orientation of the Silicon while the deviation of the peak positions relative to that of the unstrained Laue pattern determine the deviatoric strain tensor [40].

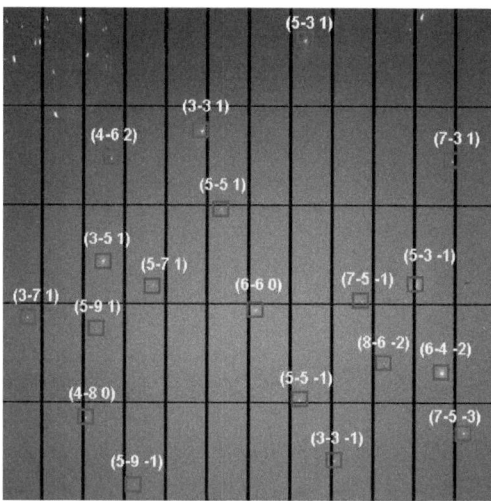

Figure 4.12 Indexed Laue microdiffraction pattern of silicon at a depth of approximately 5 μm.

Figure 4.13 illustrates the results of the 20x20 μm large stress mapping by means of synchrotron X-ray Laue microdiffraction. The maps of Figure 4.13 a) and b) present the shear stress component S_{zy} and S_{xy}, respectively. The bottom of the maps corresponds to the border of the sample. While the shear stress (see Figure 4.13 a)) is in the whole mapping in the tensile regime, the stress component S_{xy} is in the compressive state. Both shear stress components show the maximal absolute stress values in an approximately 4-6 μm wide band parallel to the x-axis. The band starts to appear circa 6 to 8 μm away from the border of the sample. With increasing distance from these bands the stresses are slowly decreasing. The maximal stresses are in the case of S_{zy} in the order of 60 MPa while the shear stress component S_{xy} shows maximal stresses of around -40 MPa.

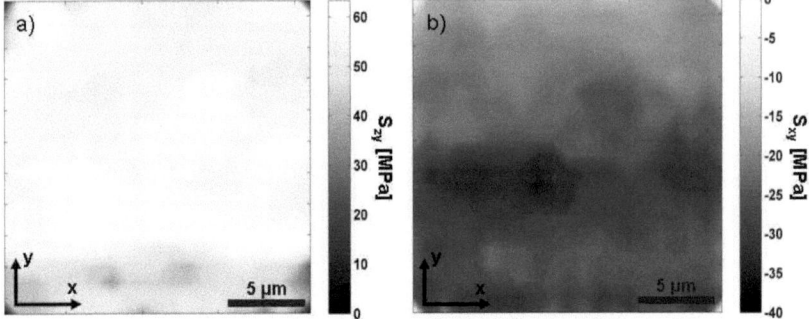

Figure 4.13 20x20 μm large mapping of two shear stress components calculated from the Laue microdiffraction pattern. a) map of the S_{zy} shear stress component. b) map of the S_{xy} shear stress component. The bottom of the maps corresponds to the border of the sample.

This band of maximal shear stresses corresponds very well with the region where the maximal silicon Raman peak shift was observed (see Figure 4.8) and the maximal density of defects and implanted He^+-ions (see Figure 4.11). Therefore, one can conclude that the measured shear stresses appear due the implanted He^+-ions. As already mentioned, it was not possible to measure the residual stresses by means of Raman microscopy as the implanted He^+-ions cause a phonon confinement which also influences the Raman peak position. In principle, the phonon confinement leads not only to a shift of the Raman peak but also to an increase of the asymmetry of the peak. If it is possible to calculate the expected Raman peak shift by analyzing the asymmetry of the Raman peak, then it would be possible to separate the component of the peak shift caused by residual stresses and the component of the peak shift due to phonon confinement.

4.8.2 Si^{2+}-ion Implantation in Silicon

This section contains additional experiments and results regarding the influence of implanted ions on the Raman spectrum of silicon. Pieces of a silicon wafer were irradiated with Si^{2+}-ions at three different fluences of $1.5*10^{15}$, $2.0*10^{16}$ and $3.5*10^{16}$ Si^{2+}/cm^2. The ions were implanted with an energy of 4 MeV. Si^{2+}-ions (self ions) were chosen to avoid any alloying effects which could have an influence on the Raman spectrum. The irradiation direction was inclined by 8° to the surface normal to avoid channeling. After the irradiation, the silicon wafer was cracked through the center of the irradiated area. The cracked surface was mapped by the confocal Raman microscope (see Figure 4.2). After an annealing for 1 hour at 700°C and at a pressure of 10^{-6} mbar, the samples where mapped again by means of Raman microscopy to see whether the defects are able to recover or not.

Figure 4.14 illustrates the result of the maps of the cracked Si-wafer. The map in Figure 4.14 a) is a non-irradiated Si-wafer which was used as a reference measurement. Such a measurement was necessary as the cracking of the wafers might induce phase transformations or residual stresses which would both influence the Raman measurements. The maps b), c) and d) in Figure 4.14 are the irradiated samples with a fluence of $1.5*10^{15}$, $2.0*10^{16}$ and $3.5*10^{16}$ Si^{2+}/cm^2, respectively. "A" corresponds in all maps to the area where the Raman signal is not shifted with respect to the pristine Si Raman signal. "B" marks the region where the peak position is shifted to lower wavenumbers. "C" marks the border of the sample. All the irradiated samples show directly at the border an approximately 4 µm wide zone where the peak position is shifted to lower wavenumbers. Such a region is not visible in the non-irradiated Si sample. The maps show that with increasing dose of implanted ions the peak shift increases. The map of the non-irradiated sample in Figure 4.14 a) showed that cracking neither causes residual stresses nor phase transformations to a larger extent. In map b) the region "B" is shifted 0.5 cm^{-1} with respect to the normal peak position. In map c) the shift is in the order of $0.5 - 0.8$ cm^{-1}. In the map d) the area "B" can be divided further into two different zones. In the first region, which is directly at the border of the cracked wafer, the silicon Raman peak is shifted $0.5 - 1.0$ cm^{-1}. In the second region, 2-4 μm away from the border of the Si-wafer, the peak position is shifted down to 470 cm^{-1}. The spectra which have a peak position at around 470 cm^{-1} clearly belong to amorphous silicon (see Figure 4.6) while spectra having peak shifts up to 1.0 cm^{-1} origin from crystalline silicon. As in the case for He^{+}-ion implantation, the behavior can be explained by the phonon confinement model (PCM). It assumes that phonons are not infinite anymore as in a single crystal but constrained by grain boundaries. The restriction causes a relaxation of the phonon wave vector. An increasing peak shift corresponds to a decreasing grain size.

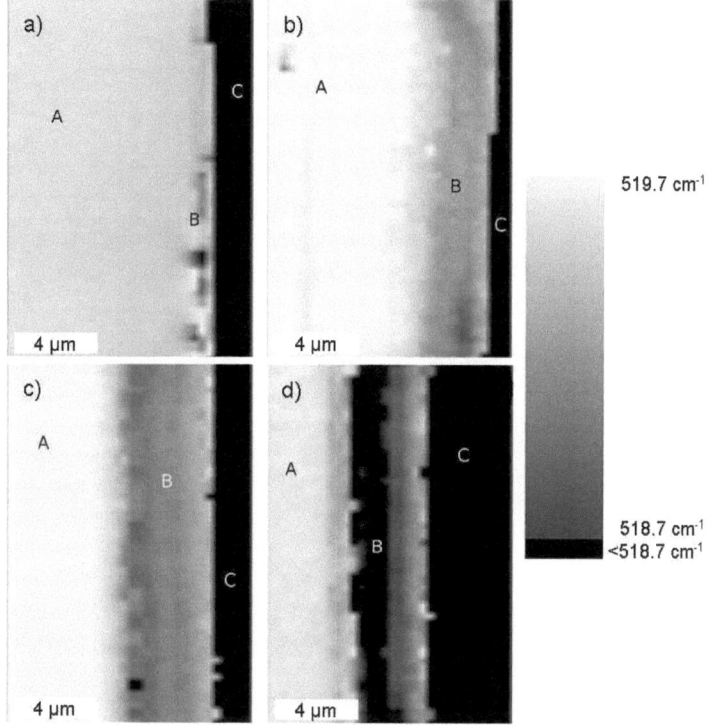

Figure 4.14 Raman maps of the cracked surface of a) a non-irradiated Si-wafer, b) the Si-wafer irradiated with a fluence of $1.5*10^{15}$ Si^{2+}/cm^2, c) irradiated with a fluence of $2.0*10^{16}$ Si^{2+}/cm^2 and d) with fluence of $3.5*10^{16}$ Si^{2+}/cm^2. "A" corresponds in all maps to the area where the Raman peak is not shifted. "B" belongs to the area where the Si-peak is shifted to lower wavenumbers. "C" marks the border of the samples.

Like in the experiments of the He^+-ion implantation (see Figure 4.11), a SRIM calculation was performed. The implanted ions cause defects in the crystal as they collide with the crystal atoms. Figure 4.15 a) illustrates the calculated amount of collisions events at a certain depth. The number of collisions directly correlates to the number of defects. The maximum of collisions is at a depth of 2-3 µm. Figure 4.15 b) illustrates another sort of disorder which is the stopping of the penetrating ions. The SRIM simulation predicts that great majority of atoms stop in a depth of 2.5 to 3 μm. The Raman maps show a strong correlation between the Raman peak shift and the SRIM simulation. The maximum of defects and implanted ions, therefore the area where the undamaged crystalline regions are the smallest, are at the same depth as the maximum of the peak shift in the Raman scans. The region of the amorphous silicon in Figure 4.14 d) appears at the depth where most of the defects are caused and where the ions stop within the material. As in the case for He$^+$-ion implantation, the behavior can be explained by the phonon confinement model (PCM). It assumes that phonons are not infinite anymore as in a single crystal but constrained by grain boundaries. The restriction causes a relaxation of the phonon wave vector. An increasing peak shift corresponds to a decreasing grain size.

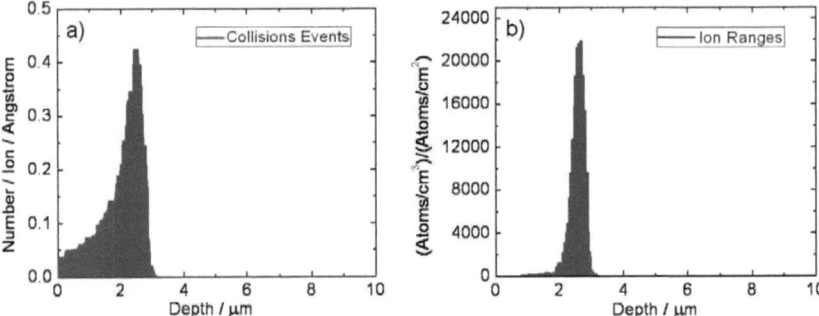

Figure 4.15 SRIM simulation[29] for the irradiation of Si^{2+} ions with an energy of 4 MeV into silicon. a) numbers of defects caused by the irradiation with a maximum at a depth of around 2 to 3 μm. b) stopping range of the Si^{2+} ions in the silicon. Most of the ions stop between 2.5 - 3 μm within the silicon.

The irradiated samples were annealed for 1 h at a temperature of 700°C in vacuum. The results are presented in Figure 4.16. The maps a), b) and c) correspond to the samples irradiated with a fluence of $1.5*10^{15}$, $2.0*10^{16}$ and $3.5*10^{16}$ Si^{2+}/cm², respectively. The symbol "A" belongs to the area where no peak shift was observed. "B" highlights spots and areas where a significant peak shift was measured while "C" marks the border of the silicon samples. The results show that the bands of lowered wavenumbers which were caused by the implantation of Si^{2+}-ions disappeared after the annealing completely. Even the band of amorphous silicon which occurred at the highest fluence dissolved (compare Figure 4.14 d) and Figure 4.16 c)). The reason is a re-crystallization of the silicon during the annealing process. The spots marked by label "B" are most probably artifacts of the cracking as similar patterns also appear in the non-irradiated sample (Figure 4.14 a)).

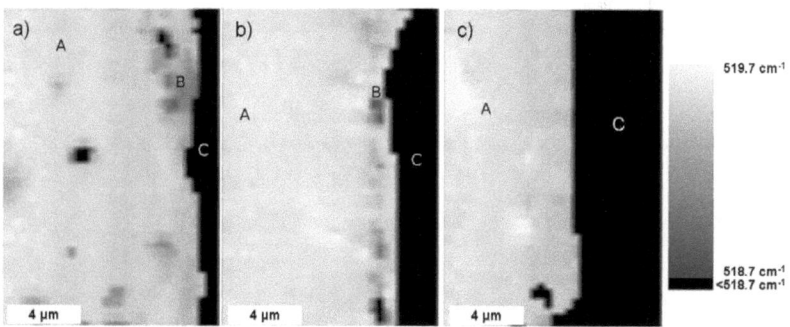

Figure 4.16 Raman maps of the cracked Si-surface after the annealing for 1h at 700°C of a) the Si-wafer irradiated with a fluence of $1.5*10^{15}$ Si^{2+}/cm^2, b) irradiated with a fluence of $2.0*10^{16}$ Si^{2+}/cm^2 and c) irradiated with fluence of $3.5*1016$ Si^{2+}/cm^2. "A" marks the zone of non-shifted Raman peaks. "B" highlights spots where the peak is shifted significantly. "C" shows the border of the silicon samples.

The experiments show that defects caused by the implantation of ions affect the Raman signal. The resulting Raman shift is not due to any kind of alloying effects. The heating of the samples leads to a recovery and to a re-crystallization of the disturbed areas. Further experiments could investigate to what degree the amorphous areas re-crystallize and how the microstructure looks like.

4.9 References

1. A. Kailer, Y.G. Gogotsi, and K.G. Nickel, *Phase transformations of silicon caused by contact loading*. Journal of Applied Physics, 1997. **81**(7): p. 3057-3063.
2. G. Weill, J.L. Mansot, G. Sagon, C. Carlone, and J.M. Besson, *Characterization of Si-III and Si-IV, metastable forms of silicon at ambient pressure*. Semiconductor Science and Technology, 1989. **4**(4): p. 280-282.
3. S.R. Jian, *Mechanical deformation induced in Si and GaN under Berkovich nanoindentation*. Nanoscale Research Letters, 2008. **3**(1): p. 6-13.
4. Y. Gogotsi, T. Miletich, M. Gardner, and M. Rosenberg, *Microindentation device for in situ study of pressure-induced phase transformations*. Review Of Scientific Instruments, 1999. **70**(12): p. 4612-4617.
5. R. Gassilloud, C. Ballif, P. Gasser, G. Buerki, and J. Michler, *Deformation mechanisms of silicon during nanoscratching*. Physica Status Solidi A-Applications and Materials Science, 2005. **202**(15): p. 2858-2869.
6. B. Moser, K. Wasmer, L. Barbieri, and J. Michler, *Strength and fracture of Si micropillars: A new scanning electron microscopy-based micro-compression test*. Journal of Materials Research, 2007. **22**(4): p. 1004-1011.
7. K. Wasmer, T. Wermelinger, A. Bidiville, R. Spolenak, and J. Michler, *In situ compression tests on micron-sized silicon pillars by Raman microscopy—Stress measurements and deformation analysis*. Journal of Materials Research, 2008. **23**(11): p. 3040-3047.
8. J. Deneen, W.M. Mook, A. Minor, W.W. Gerberich, and C.B. Carter, *In situ deformation of silicon nanospheres*. Journal of Materials Science, 2006. **41**(14): p. 4477-4483.
9. S. Rubanov and P.R. Munroe, *FIB-induced damage in silicon*. Journal of Microscopy-Oxford, 2004. **214**: p. 213-221.
10. N.I. Kato, *Reducing focused ion beam damage to transmission electron microscopy samples*. Journal of Electron Microscopy, 2004. **53**(5): p. 451-458.

11. R.M. Langford, *Focused ion beams techniques for nanomaterials characterization*. Microscopy Research and Technique, 2006. **69**(7): p. 538-549.
12. T. Ishitani and T. Yaguchi, *Cross-sectional sample preparation by focused ion beam: A review of ion-sample interaction*. Microscopy Research and Technique, 1996. **35**(4): p. 320-333.
13. J.C. Reiner, P. Nellen, and U. Sennhauser, *Gallium artefacts on FIB-milled silicon samples*. Microelectronics Reliability, 2004. **44**(9-11): p. 1583-1588.
14. T. Ishitani, H. Koike, T. Yaguchi, and T. Kamino, *Implanted gallium ion concentrations of focused-ion-beam prepared cross sections*. Journal of Vacuum Science & Technology B, 1998. **16**(4): p. 1907-1913.
15. M. Tamura, S. Shukuri, M. Moniwa, and M. Default, *Focused ion-beam gallium implantation into silicon*. Applied Physics a-Materials Science & Processing, 1986. **39**(3): p. 183-190.
16. J.C. Bourgoin, J.F. Morhange, and R. Beserman, *On amorphous layer formation in silicon by ion implantation*. Radiation Effects, 1974. **22**(3): p. 205-8.
17. C.A. Volkert and E.T. Lilleodden, *Size effects in the deformation of sub-micron Au columns*. Philosophical Magazine, 2006. **86**(33-35): p. 5567-5579.
18. D. Kiener, W. Grosinger, G. Dehm, and R. Pippan, *A further step towards an understanding of size-dependent crystal plasticity: In situ tension experiments of miniaturized single-crystal copper samples*. Acta Materialia, 2008. **56**(3): p. 580-592.
19. W.D. Nix, J.R. Greer, G. Feng, and E.T. Lilleodden, *Deformation at the nanometer and micrometer length scales: Effects of strain gradients and dislocation starvation*. Thin Solid Films, 2007. **515**(6): p. 3152-3157.
20. M.D. Uchic and D.A. Dimiduk, *A methodology to investigate size scale effects in crystalline plasticity using uniaxial compression testing*. Materials Science and Engineering a-Structural Materials Properties Microstructure and Processing, 2005. **400**: p. 268-278.
21. H. Bei and S. Shim, *Effects of focused ion beam milling on the nanomechanical behavior of a molybdenum-alloy single crystal*. Applied Physics Letters, 2007. **91**.

22. Y. Gogotsi, C. Baek, and F. Kirscht, *Raman microspectroscopy study of processing-induced phase transformations and residual stress in silicon.* Semiconductor Science and Technology, 1999. **14**(10): p. 936-944.
23. I. De Wolf, *Micro-Raman spectroscopy to study local mechanical stress in silicon integrated circuits.* Semiconductor Science and Technology, 1996. **11**(2): p. 139-154.
24. Z. Iqbal and S. Veprek, *Raman-scattering rom hydrogenated microcrystalline and amorphous-silicon.* Journal of Physics C-Solid State Physics, 1982. **15**(2): p. 377-392.
25. R. Ossikovski, Q. Nguyen, G. Picardi, J. Schreiber, and P. Morin, *Theory and experiment of large numerical aperture objective Raman microscopy: application to the stress-tensor determination in strained cubic materials.* Journal of Raman Spectroscopy, 2008. **39**(5): p. 661-672.
26. E. Bonera, M. Fanciulli, and D.N. Batchelder, *Combining high resolution and tensorial analysis in Raman stress measurements of silicon.* Journal of Applied Physics, 2003. **94**(4): p. 2729-2740.
27. T. Wermelinger, C. Borgia, C. Solenthaler, and R. Spolenak, *3-D Raman spectroscopy measurements of the symmetry of residual stress fields in plastically deformed sapphire crystals.* Acta Materialia, 2007. **55**(14): p. 4657-4665.
28. J. Comas and R.G. Wilson, *Channeling and random eqvalent depth distribution of 150 keV Li, Be and B implanted in Si.* Journal of Applied Physics, 1980. **51**(7): p. 3697-3701.
29. J.F. Ziegler, J.P. Biersack, and U. Littmark, *Stopping and range of ions in solids*, 321, Editor. 1985.
30. V.D. Blank and B.A. Kulnitskiy, *Crystallogeometry of polymorphic transitions in silicon under pressure.* High Pressure Research, 1996. **15**(1): p. 31-42.
31. X. Huang, F. Ninio, L.F. Brown, and S. Prawer, *Raman-scattering studeis of surface modification in 1.5 MeV Si-implanted silicon.* Journal of Applied Physics, 1995. **77**(11): p. 5910-5915.
32. I. De Wolf, *Micro-Raman spectroscopy to study local mechanical stress in silicon integrated circuits.* Semiconductor science and technology 1995. **11**: p. 139-154.

33. K.K. Tiong, P.M. Amirtharaj, F.H. Pollak, and D.E. Aspnes, *Effects of As^+ ion implantation on the Raman-spectra of GaAs - spatial correlation interpretation.* Applied Physics Letters, 1984. **44**(1): p. 122-124.
34. H. Richter, Z.P. Wang, and L. Ley, *The one phonon Raman-spectrum in microcrystalline silicon.* Solid State Communications, 1981. **39**(5): p. 625-629.
35. H. Tang and I.P. Herman, *Raman microprobe scattering of solid silicon and germanium at the melting temperature.* Physical Review B, 1991. **43**(3): p. 2299-2304.
36. R. Siegele, G.C. Weatherly, H.K. Haugen, D.J. Lockwood, and L.M. Howe, *Helium Bubbles in Silicon - Structure and Optical-Properties.* Applied Physics Letters, 1995. **66**(11): p. 1319-1321.
37. A.A. MacDowell, R.S. Celestre, N. Tamura, R. Spolenak, B. Valek, W.L. Brown, J.C. Bravman, H.A. Padmore, B.W. Batterman, and J.R. Patel, *Submicron X-ray diffraction.* Nuclear Instruments & Methods In Physics Research Section A-Accelerators Spectrometers Detectors And Associated Equipment, 2001. **467**: p. 936-943.
38. Brönnimann, *The Pilatus 1M Detector.* Journal of Synchrotron Radiation, 2006. **13**: p. 120-130.
39. T. Weber, S. Deloudi, M. Kobas, Y. Yokoyama, A. Inoue, and W. Steurer, *Reciprocal-space imaging of a real quasicrystal. A feasibility study with PILATUS 6M.* Journal of Applied Crystallography, 2008. **41**(4): p. 669-674.
40. N. Tamura, A.A. MacDowell, R. Spolenak, B.C. Valek, J.C. Bravman, W.L. Brown, R.S. Celestre, H.A. Padmore, B.W. Batterman, and J.R. Patel, *Scanning X-ray microdiffraction with submicrometer white beam for strain/stress and orientation mapping in thin films.* Journal of Synchrotron Radiation, 2003. 10: p. 137-143.

5 3-D Raman Spectroscopy Measurements of the Symmetry of Residual Stress Fields in Plastically Deformed Sapphire Crystals

This chapter illustrates how confocal Raman microscopy can be used for analyzing stress fields in three dimensions with a spatial resolution in the submicrometer range. For this study sapphire was chosen as it is transparent as well as Raman active. The results show very nicely how the residual 3-D stress fields around a micro indent correlates to the deformation mechanism of the material rather than to the geometry of the indent itself. A drawback of the method is that the number of components of the stress tensor which are accessible depends directly on the material. As 3-D stress states are normally very complicated the results show a more qualitatively overview. For quantitative results one has to assume simplified stress states. This work is published in a full length article:
Wermelinger T, Borgia C., Solenthaler C., Spolenak R; *3-D Raman spectroscopy measurements of the symmetry of residual stress fields in plastically deformed sapphire crystals,* Acta Materialia 55 (2007) 4657-4665

5.1 Abstract

Methods for measuring stresses in three dimensions have recently received high attention. Specifically the stress fields around indents in metals were studied by 3-D X-ray stress microscopy. In this paper we investigate the 3-D residual stress field around a microindent using confocal Raman microscopy with a lateral resolution of 300 nm and a depth resolution of 600 nm. The model system investigated was single crystalline sapphire, which was indented normal to its basal c(0001) plane. A cross section of the indent was studied by TEM to visualize defect structures. The major result is that in sapphire the geometry of the indenter has no direct influence on the symmetry of the resulting residual stress field. Residual stresses directly depend on the crystal symmetry of the crystal and the defect structures within. Confocal Raman microscopy is a powerful method for analyzing 3-D stress fields and the corresponding defect structures by peak width analysis with a resolution in the submicron range.

5.2 Introduction

Novel applications in the microelectronics industry have renewed the interest in sapphire as a substrate [1]. Various sapphire based devices such as fast neutron filters and high pressure magnetic resonance cells have been developed. Specifically its high strength and heat resistance of sapphire attracts a lot of attention. Therefore a deep understanding of the mechanical properties is required.

In this context the deformation of sapphire has been extensively studied by several groups [1-12]. As a result, the deformation and fracture mechanisms of the sapphire crystal are well known. In contrast, magnitude and distribution of residual internal stresses after plastic deformation, i.e. the properties of residual stress fields, still remain unclear. Since these stresses may cause failure, hence affect lifetime of devices, it is obviously important to measure and to visualize them, ideally in three dimensions (3-D). At the moment, only few methods are known to perform 3-D measurements with a resolution in the micron range.

Currently only X-ray micro beam diffraction is an experimental option to measure the strain tensor with submicron resolution [13-15] in three dimension. The 3-D distribution of the local crystalline phase, texture and the elastic strain tensor can be measured with a resolution below 1 μm. These instruments combine ultra intense synchrotron X-ray sources and advanced X-ray optics to probe crystalline materials. Only a few synchrotron sources are available world wide, and until now only a few of these experiments have been performed therefore other methods on the laboratory scale need to be established.

This paper demonstrates that another accurate method for the 3-D measurements of stresses is micro Raman spectroscopy. The method is based on the well known effect that the Raman signal (phonon energies) may be affected by internal stresses. Therefore, in a Raman spectrum of deformed material, peak positions are shifted away from the peak positions obtained from the stress free material. Hence, quantifying this shift allows determining sign and magnitude of internal stress. As already proven in many studies for the 2-D case [16-20], Raman microscopy was successfully applied for analyzing stresses in silicon devices and other Raman active materials such as polymers, semiconductors and ceramics. In transparent Raman active materials this method could also be used for the 3-D case if a confocal microscope is used that provides precise positioning of the focus spot. Such measurements at a micro indent in a sapphire single crystal are reported and discussed below.

5.2.1 Raman Effect of Sapphire

Sapphire has a rombohedral crystal structure belonging to the D_{3d}^6 space group with two molecular Al_2O_3 groups per unit cell [21, 22]. This leads to seven Raman-active phonon modes: $2A_{1g} + 5E_g$. According to Louden [23] the Raman tensors of the active optical vibrations of sapphire are:

$$A_{1g} = \begin{vmatrix} a & 0 & 0 \\ 0 & a & 0 \\ 0 & 0 & b \end{vmatrix}, \quad E_g = \begin{vmatrix} c & -c & -d \\ -c & c & d \\ -d & d & 0 \end{vmatrix} \quad (5.1)$$

The Raman frequencies (relative wave numbers) of stress free sapphire are 417 and 646 cm^{-1} for the A_{1g} modes and 380, 432, 451, 578 and 751 cm^{-1} for the E_g modes [21]. It is well known, that the phonon frequencies are shifted due to applied pressure [24, 25]. Watson *et al.* applied uniaxial pressure up to 1 GPa on macroscopic sapphire pillars and measured the corresponding peak shifts [22]. The results of these experiments are listed in Table 5.1. The same experiment but with sapphire fibers with a diameter in the micrometer range, showed slightly different results [26]. Here, the shifts were found to be about 1.8 cm^{-1}/GPa and 1.1 cm^{-1}/GPa for the peaks at 417 cm^{-1} and 380 cm^{-1}, respectively. Gallas *et al.*[25] used a diamond anvil high pressure cell (DAC) for the calibration of the Raman effect in sapphire. They received similar results for the peak at 380 cm^{-1}, however for the peak at a relative wavelength of 417 cm^{-1} they found a pressure dependence of 2.2 cm^{-1}/GPa. Shin *et al.* reported the uniaxial stress on the Raman frequencies in sapphire [27].

Table 5.1 Pressure dependence of Raman frequencies in sapphire.

Frequency (cm^{-1})	Peak shift (cm^{-1}/GPa)			
	Shin *et al* [27].	Watson *et al.*[22]	Gallas *et al.*[25]	Jia and Yen [26]
380	2.3 ± 0.2	1.37 ± 0.06	1.10 ± 0.10	1.10
417	1.7 ± 0.1	2.11 ± 0.06	2.20 ± 0.07	1.80
432	1.8 ± 0.1	2.95 ± 0.08	-	1.16
451	1.0 ± 0.2	1.66 ± 0.10	-	-
578	2.7 ± 0.3	2.77 ± 0.12	-	0.55
646	5.0 ± 0.4	-	-	0.00
751	2.5 ± 0.3	4.80 ± 0.20	-	1.70

5.2.2 Experimental

A sapphire single crystal (size 10x10x1 mm) was plastically deformed by micro indentation. The indent was placed on the basal (0001) plane with a Vickers micro indenter at room temperature. A maximal indentation force of 0.4 N at ambient pressure was applied under constant load rate of 0.05 N/sec. The radius of the indenter tip varied from one direction to the other from 0.16 µm to 1.1 µm. The shape of the indent was mapped with an atomic force microscope (AFM) (CRAFM 200, WITec GmbH, Germany).

The residual stress field around the indent was measured with a confocal Raman microscope (CRM 200, WITec GmbH, Germany) equipped with a Helium-Cadmium laser with a wavelength of 442 nm. The microscope was operated in the backscattering mode with a 100x objective and 0.9 numerical apperture. According the Rayleigh equation:

$$d = 1.22 \cdot \frac{\lambda_{Laser}}{2 \cdot NA}, \qquad (5.2)$$

the lateral resolution of the microscope of 300 nm while the depth resolution is on the order of 600 nm. As the microscope is confocal only information from the focus spot is collected. The confocality of the microscope combined with the possibility of an exact positioning of the focus spot in x-,y- and z-directions allows measuring stress fields in 3 dimensional.

Starting from the surface of the sapphire crystal, a stack of planar Raman scans parallel to the basal (0001) plane was recorded. Measurements were made always at the same x- and y-coordinates but changing the z-coordinate. The interplanar distance between two Raman scans was 0.4 μm. Totally 19 scans were taken down to 7.6 μm below the surface of the crystal. Each planar scan was 12x12 μm wide and consisted of 48x48 spectra. As the Raman signal of sapphire was weak compared to signals form silicon an integration time of 5 sec per spectrum was chosen to obtain a sufficiently large signal to noise ratio.

In order to obtain some direct information on the deformation structure, a cross section trough the indent was examined by means of transmission electron microscopy (TEM). For the purpose a TEM-lamella was cut perpendicular to the (0001) basal plane through the center of the indent (see Figure 5.2) with a FEI Nano Lab 600 Focused Ion Beam machine (FIB). The TEM-lamella was about 2.5x15 μm wide and 100 nm to 150 nm thin; it was examined with a FEI Tecnai G^2 F20X-Twin TEM operating at 200 kV.

5.3 Results

Figure 5.1 shows the Raman effect in sapphire for two different crystal orientation regarding the polarization of the incoming laser light. The black spectrum (rectangular symbols) was taken with an incident laser beam polarized perpendicular to the c(0001) axis. In this orientation six major lines appear at wavenumbers of 378, 417, 428, 447, 575 and 748 cm^{-1}. The gray spectrum (triangular symbols) was taken with an incident laser beam polarized parallel to the c(0001) axis and shows three prominent lines at 379, 417, 645 cm^{-1} and a faint one at 748 cm^{-1}. All these results are in agreement with the literature [21] results. The peak at 378 cm^{-1} gets bigger in comparison to the other peaks with increasing numerical aperture of the objective. This leads to the conclusion that this vibration has a bigger in plane component than the other phonons due to the fact that a parallel laser beam only interacts with out of plane phonons.

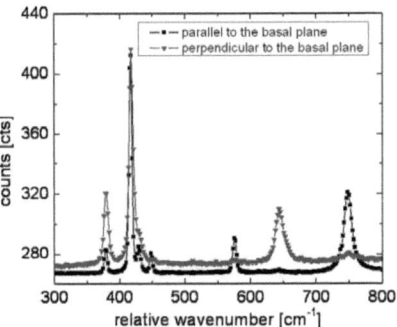

Figure 5.1 Raman spectra of sapphire single crystal. The black spectrum (rectangular symbols) was taken with an incident laser light polarized perpendicular to the c(0001) axis. The gray spectrum (triangular symbols) was taken with an incident laser light polarized parallel to the c(0001) axis. The spectra were taken with a 100x objective with an numerical aperture of 0.9.

Figure 5.2 a) shows the topography of the Vickers microindent on the sapphire single crystal on the basal c(0001) plane mapped with an AFM. The Vickers indent has a diameter of about 4.5 to 5 µm. No pile-ups are visible at the edges of the indent. Although there are some slightly asymmetric features such as the tip radius, it is very obvious that the plastically deformed area shows the four-fold symmetry of the indenter tip. A cross-section through the middle of the indent shows that the indent has a depth on the order of 0.5 µm.

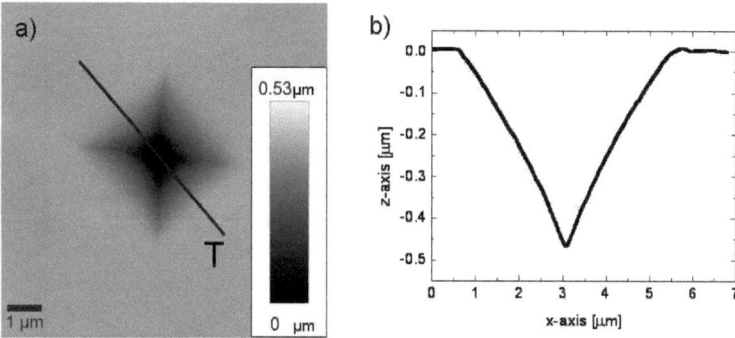

Figure 5.2 a) shows the AFM image of the Vickers Microindent in the basal (0001) plane of the sapphire single crystal. The indent has a diameter of about 4.5 to 5 µm and shows clearly a four fold symmetry. b) is a crossection through the middle of the indent which has a depth of approximately 0.5 µm.

All measured peaks show shifts due to residual stresses. Figure 5.3 shows the peak shift of four different peaks 0.4 µm below the surface of the sample. While a shift to higher wavenumbers stands for compressive stresses a shift to lower wavenumbers represents tensile stresses. The stress maps b)-d) show the stress distribution in the out-of-plane direction. While the stress map a) consists of mixture between in-plane and put-of-plane stresses. Although all the peaks show the same stress pattern: a three-fold symmetry; the clarity of the pattern strongly depends on the signal to noise ratio of the peaks occurring at different wave patterns.

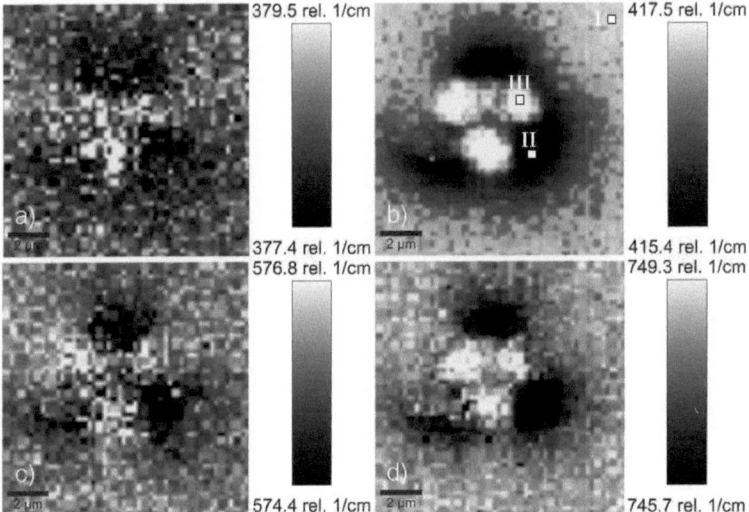

Figure 5.3 12×12 μm stress maps around the indent at 0.4 μm below the surface for the peaks at a) 378 cm^{-1}, b) 417 cm^{-1}, c) 575 cm^{-1} and d) 748 cm^{-1} wavenumbers. In the maps b)-d) bright colors correspond to compressive stresses while dark colors correspond to tensile stresses in the out-of-plane direction. Map a) shows a stress map which contains a in-plane as well as a out-of-plane component.

From Figure 5.3 it is obvious that the peak at the wavenumber at 417 cm^{-1} has the best signal to noise ratio and therefore gives the images with the lowest scattering. Figure 5.4 shows the peak shift around the indent starting from the surface of the crystal down to 3.6 μm below the surface of the sample. Bright colors correspond to compressive stresses while dark colors correspond to tensile stresses in the out-of-plane direction. On the surface one can vaguely discern the four-fold symmetry of the indent, but as soon as one focuses into the crystal this symmetry is vanishing and a three-fold symmetry is appearing. The measurements were made down to 7.6 μm but the three-fold symmetry is only visible down to 3.2 μm. As a result one obtains a stack of stress maps of the same lateral position but with changing z-coordinate.

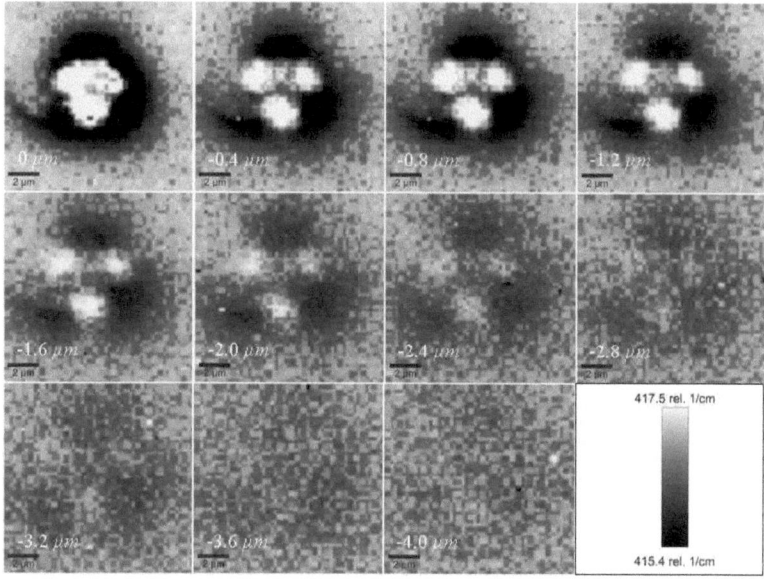

Figure 5.4 Shift of the peak at 417 cm^{-1} wavenumbers around the indent. High wavenumbers correspond to compressive stress while low wavenumbers correspond to tensile stress. The interplane distance is 0.4 µm.

Image processing tools such as ImageG give the chance to calculate from the stack of 2 dimensional maps the 3 dimensional images. Figure 5.5 shows the 3-D stress field around the Vickers indent calculated from the stack of images in Figure 5.4. In this case out-of-plane stresses are shown. As every color represents a certain stress image processing makes it possible to calculate the volume of several stress components. In principal also the volume of phase transformed regions could be determined. However, in this case the volume of this particular area was found to be too small for quantification.

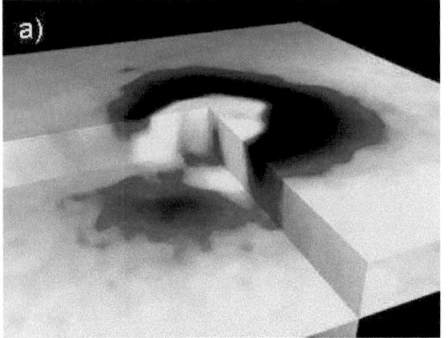 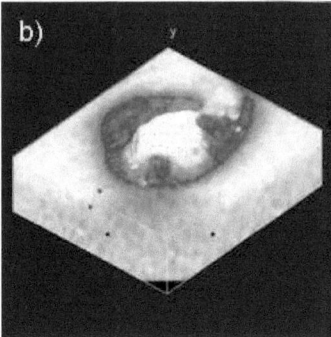

Figure 5.5 a) The calculated 3-D stress field (20x20x2 μm) around the micro indent. Dark colors refer to tensile stresses wheras bright colors correspond to compressive stresses. b) The tensile stresses above a certain threshold are transparent. Only areas with neutral or compressive stresses are visible in this volume (12x12x4 μm).

As a further result of the stacking of images one can extract single line depth scans. On the graph in Figure 5.6 a) three line scans in z-direction at different position are shown. The peak position of the peak at 417 cm^{-1} is show as a function of the depth. One scan was taken far away form the indent (Figure 5.1 arrow I). The other two scans were taken in the indented area, one of them in a region were tensile stresses were dominating (Figure 5.1 arrow II) and the other in an area with compressive stress (Figure 5.1 arrow III). The position was determined by fitting the intensity to a Gaussian function. Therefore the error bars are representative of to the standard deviation of the fitted curve. At a depth of about 2 μm the differences of the peak position between theee lines nearly vanish. Down to 4 μm the curve in the tensile region seems to be slightly below the other two curves. The line scan taken far away from the indent shows a small shift to lower wavenumbers with increasing depth.

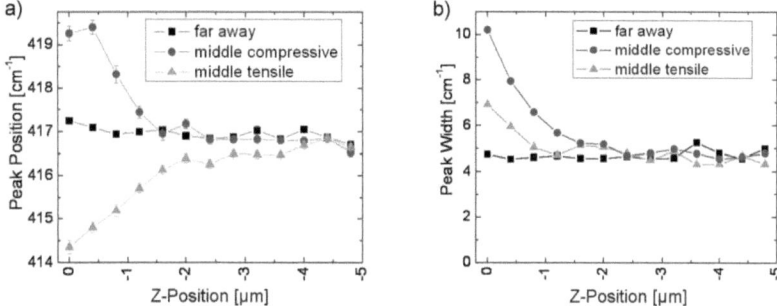

Figure 5.6 a) Peak position of three different line scans in z-direction. Taken in the unstressed part of the sample (see Figure 5.3 arrow I), in the middle of the indent where tensile stresses are dominating (Figure 5.3 arrow II) and in the middle of the indent where compressive stresses are prominent (Figure 5.3 arrow III). b) shows the exact same scans but the peak width instead of the position.

Figure 5.6 b) shows the same line scans as Figure 5.6 a) but in this case the peak width at 417 cm^{-1} is plotted as a function of the z-position. As before the major differences are visible down to a depth of about 2 μm. Interestingly, although the absolute value of the peak shift for the tensile and the compressive stress are comparable there are significant differences in peak width. As the peak width is not only influenced by the stress but also by the defect density, it could be assumed that in regions with compressive stress exhibit higher defect densities.

At the sidewalls of the sapphire indent the Raman spectrum is changing strongly (see Figure 5.7). The peaks are broadening and the peaks positions are shifted to higher wavenumbers. For example the peak at 748 cm^{-1} is shifted to 773 cm^{-1}. This behavior is very local under the sidewalls of indent and disappears in a depth of about 1.0 μm below the surface of the sample.

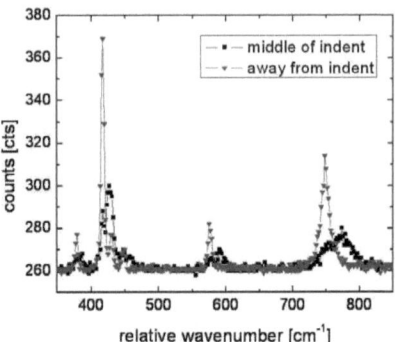

Figure 5.7 Raman spectra taken parallel to the c(0001) axis. The gray spectrum (triangles) is taken from the unstressed part of the crystal. The black spectrum (squares) is taken form the middle of the indent.

With a FIB a thin cross section through the indent was cut out. The orientation of the slice can be seen in Figure 5.2. In Figure 5.8 the TEM image of the sapphire cross section can be seen. The arrow a) points to the middle of the indent, b) shows a large deformation zone with twinning and slipping and c) marks a crack parallel to a basal slip system. The basal slip pushes material away to the sides of the indent [11]. Which slip system dominates in any region is related to the complex stress distribution around the indent. Directly under the indent twinning systems are dominating. Just below the surface the distance between single basal slip systems is small, the further away from the surface the bigger the distance between two basal slip systems. e) points to a thin layer of dislocations on the surface. It is known that polishing a sapphire induces such a thin defect layer [2]. Further there are two thin Pt-layers on the top of the sample (see arrow d)) which originate from sample preparation. Both layers were deposited to mechanically stabilize and protect the TEM lamella during the cut-out process. The first, brighter platinum layer which is directly in contact with the sapphire was deposited with an electron beam in order not to disturb the sapphire. The upper layer was deposited with an ion beam; therefore it is denser and appears darker. At the sidewalls of the indent a bright area which is marked by an ellipse f) is visible. The deformation bands which are visible in the whole deformed area do not run through this zone.

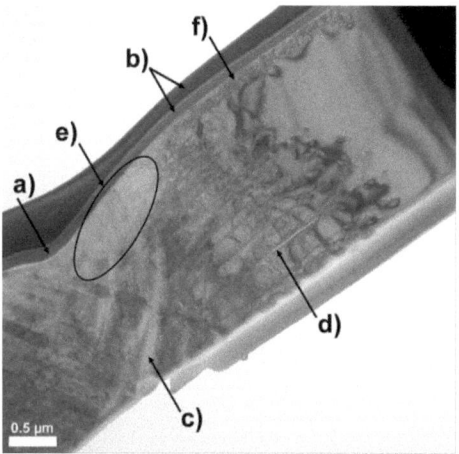

Figure 5.8 TEM image of the FIB-slice which was cut out through the indent. a) Center of the indent. b) Twinning and slipping zone. c) Crack which is parallel to the basal (0001) plane. d) A platinum layer which is needed for the sample preparation. e) Thin layer of dislocations origination from polishing. f) Highly deformed area with a possible portion of phase transformation.

5.4 Discussion

5.4.1 Mechanical Aspects of Indenting Sapphire

It is interesting to notice that the geometry of the indent has no direct influence on the symmetry of the resulting residual stress field. The residual stresses show a clear three-fold symmetry instead of the expected four-fold symmetry. The reasons for this behavior are the mechanisms of plastic deformation. To understand the appearance as well as the formation of residual stresses one has to understand the behavior of the plastic deformation. Several works investigated the deformation mechanism of sapphire single crystal at room temperature [3-7, 9, 10]. Nowak *et al.* used a spherical micro indenter for analyzing the plastic deformation. They set up a model which estimates the probability T_γ of activating the γ-th slip or twinning system as a function of the shear stress value τ_γ acting on the slip or twinning plane and the constraint factor Λ_γ denoting the orientation of the indented crystal surface, while being inversely proportional to the critical shear stress $\tau_{CR\gamma}$ [1, 5]:

$$T_\gamma = \frac{\tau_\gamma}{\tau_{CR\gamma}} \Lambda_\gamma. \qquad (5.3)$$

In hexagonal structures such as sapphire, the c/a axis ratio defines the possible slip and twinning systems. In the case of sapphire which has a c/a axis ratio of 2.73 the following deformation systems are possible:

Table 5.2. Possible twinning and slip systems of sapphire crystal [5].

Symbol	Twinning or slip system	Description	Critical shear stress (GPa)
1	$<0\bar{1}11>K1\{0\bar{1}12\}$	Rhombohedral twinning	0.111
2	$<1\bar{1}00>K1(0001)$	Basal twinning	0.148
3	$<2\bar{1}\bar{1}0>\{0\bar{1}\bar{1}2\}$	Rombohedral slip	3
4	$<2\bar{1}\bar{1}0>(0001)$	Basal slip	17
5	$<10\bar{1}0>\{\bar{1}2\bar{1}0\}$	Prismatic slip	1.2
6	$<10\bar{1}0>\{\bar{1}101\}$ $<10\bar{1}0>\{\bar{1}012\}$ $<10\bar{1}0>\{\bar{1}\bar{1}23\}$	Pyramidal slip	18

Nowak et al. calculated the probability for activating a certain slip or twining system as a function of the angle under a spherical indent [1, 5]. We assume that a spherical indent with an opening angle of 130° (180° − 2*ψ, with ψ = 25°) is comparable to the opening angle of a Vickers indent = 136°, which yields the following results. Figure 5.9 shows the possibility for activating a slip system for a spherical indentation on the c(0001) plane of sapphire for a force of 0.218 N. The angle ψ has a big influence on the deformation behavior and is one of the reasons for the appearance of a three-fold symmetry of the residual stresses. According to this calculations a flat punch indent for example would activate different deformation systems and most likely result in a six-fold symmetry in the residual stress field.

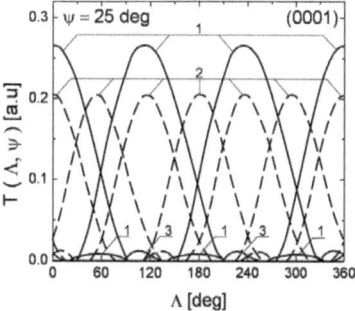

Figure 5.9 Probability for activating a slip or twinning system for all directions μ around the spherical indent [2]. The straight lines highlight the possible deformation systems in this particular case. Numbers 1-3 correspond to symbols in Table 5.2.

As the deformation state under the indent is very complex, it is not possible to say which deformation system is dominating but there seems to be some kind of cell structure. Despite the calculations performed for all slip and twinning systems in sapphire in the chosen experimental set-up the rhombohedreal twinning (1 in Figure 5.9) and the rhombohedral slip system (3 in Figure 5.9) are the dominating deformation systems. If one highlights in Figure 5.9 only these two systems, one sees immediately that the deformation shows a three fold symmetry. Due to this three fold symmetry of the deformation mechanism also residual stresses show the same behavior.

The strong changes of the Raman spectra observed directly below the indent could be due to a phase transformation. Colomban and Havel found similar peak shifts in their work [28]. They stated that, in molecular approximation, the 417 cm^{-1} band originates form an Al-O stretching mode of the AlO$_4$ tetrahedron. Therefore the shift of this peak to higher wavenumbers indicates a compressed structure. As already explained, this behavior is detectable down to 1.0 μm. If one looks at the TEM image then there is bright zone directly under the indent, where the contrast of the material is different to the other deformed area. It is to point out, that this zone is only visible on one side of the indent. This zone has a depth of about 0.5 μm. The focus spot has in z-direction a diameter of about 0.6 μm. If one compares the cut out direction of the lamella and the stress distribution, one sees that the asymmetry of the deformation structures is also visible in the stress maps. Thus it seems to be possible to measure a small signal of the phase transformed area even if the microscope is focused 1 μm below the surface, which corresponds to the findings of the measurement.

5.4.2 3-D Raman Microscopy

Tthe lateral resolution of the microscope is restricted by the wavelength of the laser and the numerical aperture. Even in the case for deep blue lasers and oil immersion objectives which have the highest numerical aperture the lateral resolution is in the order of 150 to 200 nm. In comparison with a 3-D X-ray crystal microscope [13], which has also a resolution in the submicrometer range, the lateral resolution of an optimized confocal Raman microscope is slightly better. Other advantages of the confocal Raman microscope are the simple sample preparation, the fast measuring method as well as the accessibility.

Another feature is the direct measurement of several components of the stress tensor. As the different peaks belong to different phonon vibrations of the crystal it is possible to relate different peaks to certain directions of the crystal. Therefore it is possible to assign different peak shifts to different directions. But also here, some restrictions have to be taken into account. Every crystal class has different Raman active phonon vibrations thus it is not possible to say in general how many components of the stress tensor are accessible. Silicon for example exhibits a single Raman peak which belongs to triply degenerated optical phonons [29]. By examining the polarization and the direction of the scattered light, one can obtain all components of the stress tensor [30]. For sapphire analyzed in the described experimental set-up it is possible to measure an out-of-plane component from the peaks at 417 cm^{-1}, 578 cm^{-1} and 751 cm^{-1} and a mixed out-of-plane and in-plane component from the peak at 380 cm^{-1}. Therefore only the out-of-plane component could be measured directly.

As not all components of the stress tensor could be measured it was crucial that certain assumption about the stress state have to be made to calculate stresses. While the surface was expected to be stress free, below the surface of the center of the indent the stress can be assumed to be hydrostatic. The following figure (Figure 5.10) shows the calculated stresses in the center of the indent at different depth levels based on the results of Watson et al. [22]. This calculation was chosen due to the fact that all three peaks, which show out-of-plane stresses, show similar stress values, as they should.

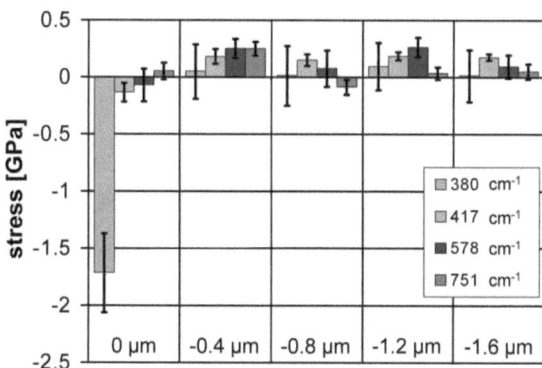

Figure 5.10 Stresses in the middle of the indent at certain depths.

At the surface of the sapphire the assumption of hydrostatic pressure is not valid. As expected the out-of-plane stress, which can be estimated from the peakshift of the peaks at 417 cm^{-1}, 578 cm^{-1} and 751 cm^{-1}, is almost zero. In contrast, the peak at a wavenumber of 380 cm^{-1} which has also an in-plane component exhibits a big compressive stress of about -1.7 GPa. Below the surface, where the pressure can be assumed hydrostatic as a rough approximation, the out-of-plane stress seems to be in the tensile regime. The peak at 380 cm^{-1} shows no significant shift which means that the tensile stress of the out-of-plane component is compensated by the in-plane stress.

Of course only transparent and Raman active materials are suitable for 3-D stress measurements. These are ceramics, diamond and most polymers. In the case of ceramics single crystal samples are preferred due to the fact that a polycrystalline sample leads to peak broadening and makes an accurate determination of the peak position more difficult. Totally amorphous samples are still Raman active but have in general such broad peaks that an exact determination of the peak position becomes impossible.

5.5 Summary and Conclusions

The residual stress distribution around an indent in sapphire was analyzed with a confocal Raman microscope. Interestingly, the residual stress field is not directly influenced by the geometry of the indent. Although the indent had four fold symmetry the residual stress field showed three fold symmetry. This result can be explained with the plastic deformation mechanisms of sapphire. Sapphire has a hexagonal crystal structure and shows therefore an anisotropic deformation. The possibility for activating a certain slip or twinning system is orientation dependent. The dominating slip and twinning system show clear three fold symmetry. This symmetry leads to a three fold symmetry of the residual stresses. Moreover, hints for a phase transformation at the sidewalls of the indent were found although the exact structure of this new phase remains unclear.

We have shown that confocal Raman microscopy is a valuable tool for analyzing 3-D stress fields with a spatial resolution in the submicron range. Depending on the Raman tensors it is possible to have access to several components of the stress tensor. In the case of sapphire only an out-of-plane component was measurable due to the fact that the experimental set-up only allowed to analyze out of plane phonons or phonons which had an out-of-plane as well as an in plane component. This method can be used for all Raman active, transparent materials.

5.6 Acknowledgments

We thank Kyburz AG for donating us a single crystal sapphire. Moreover, we thank Dr. Steve Reyntjens from FEI for the sample preparation of the TEM lamella as well as Dr. Erwan Soutry for the TEM analysis.

5.7 References

1. R. Nowak, T. Manninen, C.L. Li, K. Heiskanen, S.P. Hannula, V. Lindroos, T. Soga, and F. Yoshida, *Anomalous surface deformation of sapphire clarified by 3D-FEM simulation of the nanoindentation*. Jsme International Journal Series A-Solid Mechanics and Material Engineering, 2003. **46**(3): p. 265-271.
2. T.S.a.T. Hirayama, *Lattice Strain and Dislocations in Polished Surfaces on Sapphire*. Journal of The American Ceramic Society, 2005. **88**(8): p. 2277-2285.
3. P. Pirouz, B.F. Lawlor, T. Geipel, J.B. BildeSorensen, A.H. Heuer, and K.P.D. Lagerlof, *On basal slip and basal twinning in sapphire (alpha-Al2O3) .2. A new model of basal twinning*. Acta Materialia, 1996. **44**(5): p. 2153-2164.
4. R. Nowak and M. Sakai, *The Anisotropy Of Surface Deformation Of Sapphire - Continuous Indentation Of Triangular Indenter*. Acta Metallurgica et Materialia, 1994. **42**(8): p. 2879-2891.
5. R. Nowak, T. Sekino, and K. Niihara, *Surface deformation of sapphire crystal*. Philosophical Magazine A-Physics of Condensed Matter Structure Defects And Mechanical Properties, 1996. **74**(1): p. 171-194.
6. R. Nowak, T. Sekino, and K. Niihara, *Non-linear surface deformation of the (10(1)over-bar-0) plane of sapphire: Identification of the linear features around spherical impressions*. Acta Materialia, 1999. **47**(17): p. 4329-4338.
7. B.J. Inkson, *Dislocations and twinning activated by the abrasion of Al2O3*. Acta Materialia, 2000. **48**(8): p. 1883-1895.
8. J.B. BildeSorensen, B.F. Lawlor, T. Geipel, P. Pirouz, A.H. Heuer, and K.P.D. Lagerlof, *On basal slip and basal twinning in sapphire (alpha-Al2O3) .1. Basal slip revisited*. Acta Materialia, 1996. **44**(5): p. 2145-2152.
9. N.I. Tymiak, A. Daugela, T.J. Wyrobek, and O.L. Warren, *Acoustic emission monitoring of the earliest stages of contact-induced plasticity in sapphire*. Acta Materialia, 2004. **52**(3): p. 553-563.
10. T.F. Page, W.C. Oliver, and C.J. McHargue, *The Deformation-Behavior Of Ceramic Crystals Subjected To Very Low Load (Nano)Indentations*. Journal of Materials Research, 1992. **7**(2): p. 450-473.

11. S.J. Lloyd, J.M. Molina-Aldareguia, and W.J. Clegg, *Deformation under nanoindents in sapphire, spinel and magnesia examined using transmission electron microscopy*. Philosophical Magazine A-Physics of Condensed Matter Structure Defects And Mechanical Properties, 2002. **82**(10): p. 1963-1969.
12. W. Kollenberg, *Plastic-Deformation Of Al2o3 Single-Crystals By Indentation At Temperatures Up To 750-Degrees-C*. Journal of Materials Science, 1988. **23**(9): p. 3321-3325.
13. G.E. Ice and B.C. Larson, *3D X-ray crystal microscope*. Advanced Engineering Materials, 2000. **2**(10): p. 643-646.
14. B.C. Larson, W. Yang, G.E. Ice, J.D. Budai, and J.Z. Tischler, *Three-dimensional X-ray structural microscopy with submicrometre resolution*. Nature, 2002. **415**(6874): p. 887-890.
15. N. Tamura, A.A. MacDowell, R. Spolenak, B.C. Valek, J.C. Bravman, W.L. Brown, R.S. Celestre, H.A. Padmore, B.W. Batterman, and J.R. Patel, *Scanning X-ray microdiffraction with submicrometer white beam for strain/stress and orientation mapping in thin films*. Journal of Synchrotron Radiation, 2003. **10**: p. 137-143.
16. G.H. Loechelt, N.G. Cave, and J. Menendez, *Polarized off-axis Raman spectroscopy: A technique for measuring stress tensors in semiconductors*. Journal of Applied Physics, 1999. **86**(11): p. 6164-6180.
17. I. Dewolf, J. Vanhellemont, A. Romanorodriguez, H. Norstrom, and H.E. Maes, *Micro-Raman Study Of Stress-Distribution In Local Isolation Structures And Correlation With Transmission Electron-Microscopy*. Journal of Applied Physics, 1992. **71**(2): p. 898-906.
18. I. De Wolf, C. Jian, and W.M. van Spengen, *The investigation of microsystems using Raman spectroscopy*. Optics and Lasers in Engineering, 2001. **36**(2): p. 213-223.
19. E. Bonera, M. Fanciulli, and D.N. Batchelder, *Combining high resolution and tensorial analysis in Raman stress measurements of silicon*. Journal of Applied Physics, 2003. **94**(4): p. 2729-2740.

20. E. Bonera, M. Fanciulli, and D.N. Batchelder, *Raman spectroscopy for a micrometric and tensorial analysis of stress in silicon*. Applied Physics Letters, 2002. **81**(18): p. 3377-3379.
21. S.P.S. Porto and R.S. Krishnan, *Raman Effect of Corundum*. Journal of Chemical Physics, 1967. **47**(3): p. 1009-&.
22. G.H. Watson, W.B. Daniels, and C.S. Wang, *Measurements of Raman Intensities And Pressure-Dependence Of Phonon Frequencies In Sapphire*. Journal of Applied Physics, 1981. **52**(2): p. 956-958.
23. R. Loudon, *Raman Effect in Crystals*. Advances on Physics, 1964. **13**(52): p. 423-&.
24. S.H. Shin, F.H. Pollak, and P.M. Raccah, *Effects of Uniaxial Stress On Raman Frequencies Of Ti2o3 And Al2o3*. Bulletin of The American Physical Society, 1974. **19**(4): p. 536-536.
25. M.R. Gallas, Y.C. Chu, and G.J. Piermarini, *Calibration of The Raman Effect In Alpha-Al2o3 Ceramic For Residual-Stress Measurements*. Journal of Materials Research, 1995. **10**(11): p. 2817-2822.
26. W. Jia and W.M. Yen, *Raman-Scattering From Sapphire Fibers*. Journal of Raman Spectroscopy, 1989. **20**(12): p. 785-788.
27. F.H.P. S. Shin, P.M. Raccah. *Proceedings of the Third International Conference on Light Scattering in Solids*. in *Third International Conference on Light Scattering in Solids*. 1976.
28. P. Colomban and M. Havel, *Raman imaging of stress-induced phase transformation in transparent ZnSe ceramic and sapphire single crystals*. Journal Of Raman Spectroscopy, 2002. **33**(10): p. 789-795.
29. I. De Wolf and E. Anastassakis, *Stress measurements in silicon devices through Raman spectroscopy: Bridging the gap between theory and experiment (vol 79, pg 7148, 1996)*. Journal Of Applied Physics, 1999. **85**(10): p. 7484-7485.
30. G.H. Loechelt, N.G. Cave, and J. Menendez, *Measuring The Tensor Nature Of Stress In Silicon Using Polarized Off-Axis Raman-Spectroscopy*. Applied Physics Letters, 1995. **66**(26): p. 3639-3641.

6 Symmetry of Residual Stress Fields of ZnO Below an Indent Measured by 3-D Raman Spectroscopy

The chapter presents a similar study like in the previous section. Again, the 3-D residual stress field was analyzed around a micro indent. The here presented measurements were performed on a zinc oxide single crystal. This particular material was chosen as two independent components of the stress tensor could be measured by means of Raman microscopy. This allowed quantifying the biaxial stresses at the surface of the crystal. Additionally the results confirm that residual stresses are mainly influenced by the deformation mechanism of the material and not by the geometry of the indent. The article will be submitted as a full length article to the Journal of Applied Physics: Wermelinger T., Spolenak R.; *Symmetry of residual stress fields of ZnO below an indent measured by 3-D Raman spectroscopy*

6.1 Abstract

ZnO is a wide gap semiconductor with interesting properties for applications in nanoelectronics as well as nanophotonics and can be used for ultraviolet nanolasers. The optical and electrical properties of ZnO are strongly influenced by residual stresses, defects as well as microstructural changes. This work presents a detailed study of the residual stresses and the microstructure in 3-D around a Vickers microindent placed on the prism plane of a ZnO single crystal. The biaxial stress field on the surface of the indent was measured using a confocal Raman microscope. The deformed microstructure around the indent was examined by 3-D Raman and catodolumniscence measurements. Further, a cross-section extracted from the center of the indent was studied by transmission electron microscopy. The results show that the symmetry of the residual stress field on the surface depends not on the geometry of the indent but on the deformation mechanism of the crystal. The 3-D Raman measurements allow calculating the volume of high-dislocation density in the vicinity of the indent.

6.2 Introduction

Zinc oxide is a semiconductor with a wurtzite-like crystal structure and a band gap of 3.4 eV. The material attracts attentions for various applications such as piezoelectric transducers, varistors and transparent conducting thin films and is a very promising material for applications in nanoelectronics and nanophotonics [1-4]. Transparent zinc oxide nanowire transistors and thin film transistors are interesting due to the uncomplicated way of synthesis of nanostructures and the possibility for optoelectronic integration [5]. Huang *et al.* showed how zinc oxide nanowires can act as ultraviolet nanolasers with a wavelength of 385 nm [6]. Another promising application of zinc oxide is in solar cells [7]. Successive and reliable applications require the profound knowledge of mechanical properties such as the mechanical deformation behavior.

The mechanical properties were first studied by conventional Vickers hardness testing mostly performed in polycrystalline samples [8, 9]. In more recent time, nanoindentation revealed to be a powerful tool for study the mechanical properties of single crystalline ZnO [10-12]. One result of these measurements was that the major deformation mechanism caused by indents on the basal plane is slip in the pyramidal as well as the basal planes. Almost all studies concentrated on the mechanical properties of the c-axis. Therefore, little is known about the mechanical properties of the a-axis. Further, the magnitude and the distribution of residual stresses after plastic deformation as well as the distributions of defect in the vicinity of an indent are unknown. Both, residual stresses and the microstructure play an important role for applications as they directly change the optical and electrical properties of ZnO [13, 14]

Raman microscopy is known to be an accurate non-destructive method for analyzing different aspects of the microstructure such as grain size, phase transformation and residual stresses [15-18]. Moreover, a qualitative analysis of the defect density is possible [19]. A big advantage of confocal Raman microscopy is the opportunity to map transparent, Raman active samples in three dimensions with a lateral resolution in the submicrometer range [20].

In this work, confocal Raman microscopy was used to map the residual stresses as well as the defect density in 3-D in the vicinity of an indent placed on the a-axis of a ZnO single crystal. The method is based on the fact that the exact peak positions are directly depending on applied mechanical stress [21-25] and that the ZnO Raman signal alters with at high defect densities [26-28]. The results were compared to cathodolumniscence (CL) and electron backscatter diffraction (EBSD) measurements. Additionally, a cross-sectional lamella from the center of the indent was studied by means of Transmission electron microscopy (TEM).

ZnO has a wurtzite symmetry with a C_{6V} symmetry. From group theory one can predict an A_1 branch, a doubly degenerated E_1 branch and two doubly degenerated E_2 branches which are Raman-active as well as two B branches which are inactive [29]. The Raman tensors of the active optical vibrations are:

$$A_1(z) = \begin{pmatrix} a & 0 & 0 \\ 0 & a & 0 \\ 0 & 0 & b \end{pmatrix}, E_1(-x) = \begin{pmatrix} 0 & 0 & c \\ 0 & 0 & 0 \\ c & 0 & 0 \end{pmatrix}, E_1(y) = \begin{pmatrix} 0 & 0 & 0 \\ 0 & 0 & d \\ 0 & d & 0 \end{pmatrix}$$

$$E_2 = \begin{pmatrix} 0 & e & 0 \\ e & 0 & 0 \\ 0 & 0 & 0 \end{pmatrix}, E_2 = \begin{pmatrix} e & 0 & 0 \\ 0 & -e & 0 \\ 0 & 0 & 0 \end{pmatrix} \quad (6.1)$$

For the A_1 and the E_1 modes the coordinate in the parenthesis designates the direction of the phonon polarization. That is, an A_1 phonon which is polarized parallel to the c-axis and leads to a peak at 381 cm^{-1} and 574 cm^{-1}. The E_1 phonons are polarized in the xy-plane and cause peaks at 407 cm^{-1} and 583 cm^{-1} respectively while the E_2 vibrations are at 101 cm^{-1} and 438 cm^{-1}.

6.3 Experimental

Which phonon vibration gives response to the Raman specturm depends on the direction and the polarization of the incident laser beam. Figure 6.1 a) illustrates three different experimental set-ups used for acquiring Raman spectra from a zinc oxide single crystal. One spectrum was measured with an incident laser beam parallel to the c-axis of the crystal. The laser was polarized perpendicular to the c-axis (arrow 1 in Figure 6.1 a). Two spectra were measured on a prism plane. The laser beam was polarized either perpendicular (arrow 2 in Figure 6.1 a) or parallel (arrow 3 in Figure 6.1 a) to the c-axis. The measurements of the single spectra were performed with a 10x-objective to minimize components of the polarization which are not polarized as designated.

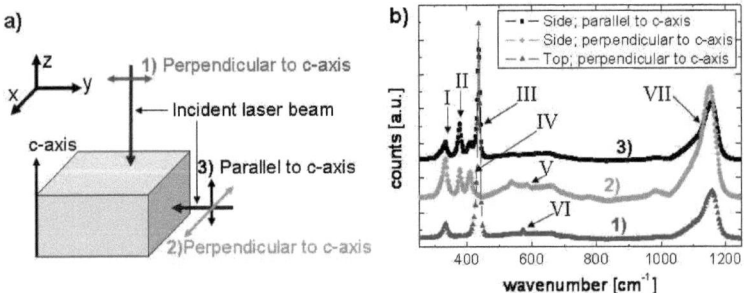

Figure 6.1. a) Schematic of the experimental set-up. Spectrum 1 was measured on the basal plane with a laser polarized perpendicular to the c-axis. The spectra measured on the prism plane were measured either with a polarization perpendicular (spectrum 2) or parallel (spectrum 3) to the c-axis. b) Raman spectra detected from the basal plane and prism plane. Peak I (at 330 cm^{-1}) and peak VII (at 1149 cm^{-1}) belong to multiphonon processes. peak II (at 378 cm^{-1}) and VI (at 574 cm^{-1}) correspond to A_1-modes and are polarized parallel to the c-axis. The peaks at 409 cm^{-1} and 583 cm^{-1} correspond to the E_1-mode and are as well as the E_2-mode peak at 438 cm^{-1} polarized parallel to the c-axis

The zinc oxide single crystal was indented with a Vickers microindenter. The indent was placed on a prism plane of the crystal at ambient conditions. The indent was performed with load of 0.1 N and with a tip having a radius varying from 0.16 μm to 1.1 μm. The area of the indent was scanned using an atomic force microscope (AFM) (CRAFM 200, WITec GmbH, Germany).

The stress field in the vicinity of indent was mapped with a confocal Raman microscope (CRM 200, WITec GmbH, Germany). The confocality of the microscope collects only scattered light from the focus spot and allows obtaining in transparent samples data from different depth levels. The microscope worked in a backscattering mode and was equipped with a helium-cadmium laser having a wavelength of 442 nm. All measurements were performed with a laser polarized perpendicular to the c-axis and a 50x-objective (numerical aperture 0.9). This leads to a lateral and depth resolution of about 0.6 μm and 1.2 μm, respectively. Starting from the surface down to a depth of 13.5 μm a stack of 2-D maps with an interplanar distance of 1.5 μm was measured. The 2-D maps had a size of 18x18 μm and consisted of 27x27 spectra. The image processing program "Image J" [30] was used to calculate 3-D structures from the stack of 2-D maps. Further, two depth scans through the center of the indent were performed having a size of 60x18 μm and consisting of 90x27 spectra. One depth scan was measured parallel to the x-axis while the other depth scan was executed along the z-axis.

Additionally, the indent was examined by means of CL and EBSD performed on a FEI Quanta 200 FEG™.

6.4 Results

The spectra illustrated in Figure 6.1 b) derive from different crystallographic orientations of the ZnO crystal (see Figure 6.1 a). Spectrum 1 was measured on the basal plane. The prominent peak at 438 cm^{-1} corresponds to an E_2-mode which is polarized perpendicular to the c-axis. The peaks I = 330 cm^{-1} and VII = 1149 cm^{-1} are due to multiphonon processes. Spectrum 2 was aquired at the prism plane with a laser polarized perpendicular to the c-axis. Peaks appear at I = 330 cm^{-1}, II = 378 cm^{-1}, III = 409 cm^{-1}, V = 583 cm^{-1}, and VII = 1149 cm^{-1}. Peak II belongs to a phonon vibration of an A_1 mode and is polarized parallel to the c-axis. The Raman peaks III and at V correspond to E_1 phonon vibrations polarized perpendicular to the c-axis [31]. Spectrum 3 was taken at the prism plane with the laser polarized parallel to the c-axis and exhibits 6 different peaks at I = 330 cm^{-1}, II = 378 cm^{-1}, III = 409 cm^{-1}, IV = 438 cm^{-1}, VI = 574 cm^{-1} and at VII = 1149 cm^{-1}. Peak VI corresponds to an A_1 phonon vibration polarized perpendicular to the c-axis.

Figure 6.2 shows the AFM map of the indent. The indent had a diameter of approximately 7 µm and a depth of about 1.3 µm. The fourfold symmetry originated from the shape of the Vickers microindenter. The scan displays a pile-up with a height of about 0.45 µm in the upper most corner of the indent.

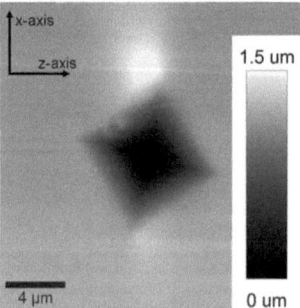

Figure 6.2. AFM scan of the Vickers microindent on the prism plane of the zinc oxide crystal. The indent has a diameter of approximately 7 µm and a depth of about 1.3 µm. The fourfold symmetry of the Vickers indent is clearly visible.

Figure 6.3 represents the shift of the peaks at a) 330 cm^{-1}, b) 378 cm^{-1}, c) 409 cm^{-1} and d) 438 cm^{-1} in the vicinity of the indent at a depth of 1.5 μm. All maps show a similar behavior, a twofold symmetry. The position of all Raman peaks is increased in a 3-6 μm wide band along the x-axis through the center of the indent. This band of increased Raman peak positions is surrounded by two much less pronounced bands where the peak position is decreased. The different signal-to-noise ratio of the different peaks influenced the clarity of the pattern. The peak at 330 cm^{-1} (Figure 6.3 a) corresponds to a multiphonon process with no polarization direction. The peak at 378 cm^{-1} (Figure 6.3 b) is polarized parallel to the c-axis. The peaks at 409 cm^{-1} (Figure 6.3 c) and 438 cm^{-1} (Figure 6.3 d) are polarized perpendicular to the c-axis of the crystal.

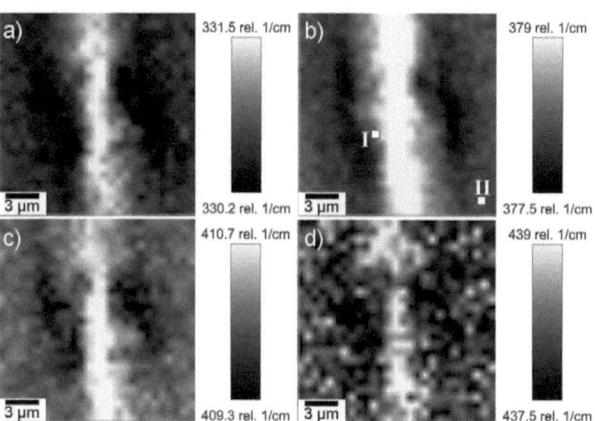

Figure 6.3. 2-D-peak position maps of the peaks at: a) 330 cm^{-1}, b) 378 cm^{-1}, c) 409 cm^{-1} and d) 438 cm^{-1} around the indent at a depth of 1.5 μm. All maps show the same pattern which has a twofold symmetry.

Figure 6.3 demonstrated the twofold symmetry pattern of the peak positions. The mapped area was too small to see the full extent of this pattern. Therefore, two depth scans through the center of the indent were performed along the x-axis and along the z-axis (Figure 6.4), respectively. In both maps the shift of the peak at 378 cm^{-1} is illustrated. The pattern of the peak position strongly differs between the two scan directions. Along the x-axis (Figure 6.4 a) different bands where the peak position is increased are visible even in a distance of 30 μm from the center of the indent. The lines in Figure 6.4 a) correspond to the direction of the bands with increased peak position. The following angles between the surface (s) and the direction of the stress band were measured: s-I = 23°, s-II = 55°, s-III = 119°, s-IV = 150° and s-V = 172°. The scan along the *c*-axis (Figure 6.4 b) shows directly below the indent down to a depth of 18 μm an increased peak position. Two small bands of decreased peak position surround the zone of the increased peak position. In both maps a maximal increase of 2 cm^{-1} was found in the center of the indent.

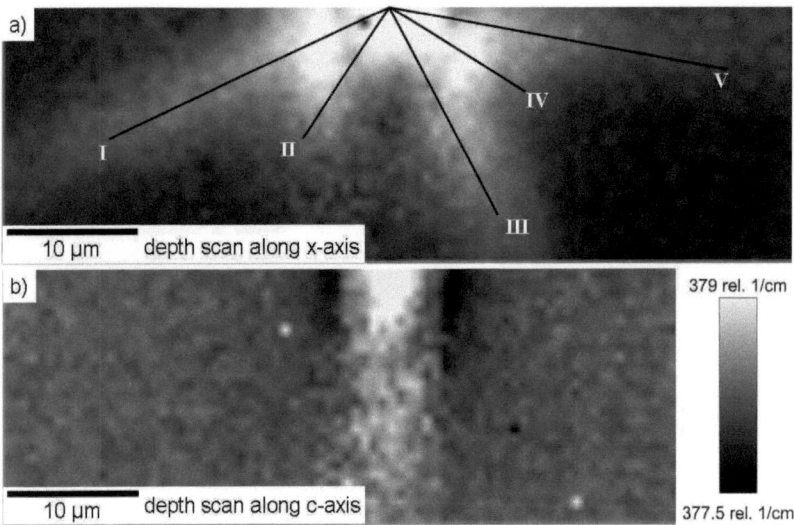

Figure 6.4. depth scans a) parallel to the x-axis or b) parallel to the c-axis. The figures show the shift of the peak at 378 cm^{-1}. The lines in a) correspond to the direction of the bands of increased peak position.

Directly below the indent a new peak at 583 cm^{-1} in the Raman spectrum was observed. Also the intensity of the peak at 438 cm^{-1} was slightly increased. The Raman spectra in Figure 6.5 derive from a pristine region of the ZnO crystal (gray triangles) from the center of the indent (black squares), respectively. Both spectra were extracted from a 2-D map (see Figure 6.3). Beside the two increased peaks no differences in terms of peak intensity as well as peak width were observed.

In Figure 6.5 the intensity of the peak at 583 cm^{-1} is presented from the surface down to 13.5 μm. With increasing depth the intensity of the peak decreases. Down to a depth of 4.5 μm the intensity distribution of the peak highly corresponds to the symmetry of the indent. In the area between 6 μm and 13.5 μm the intensity is slightly increased but no pattern is visible.

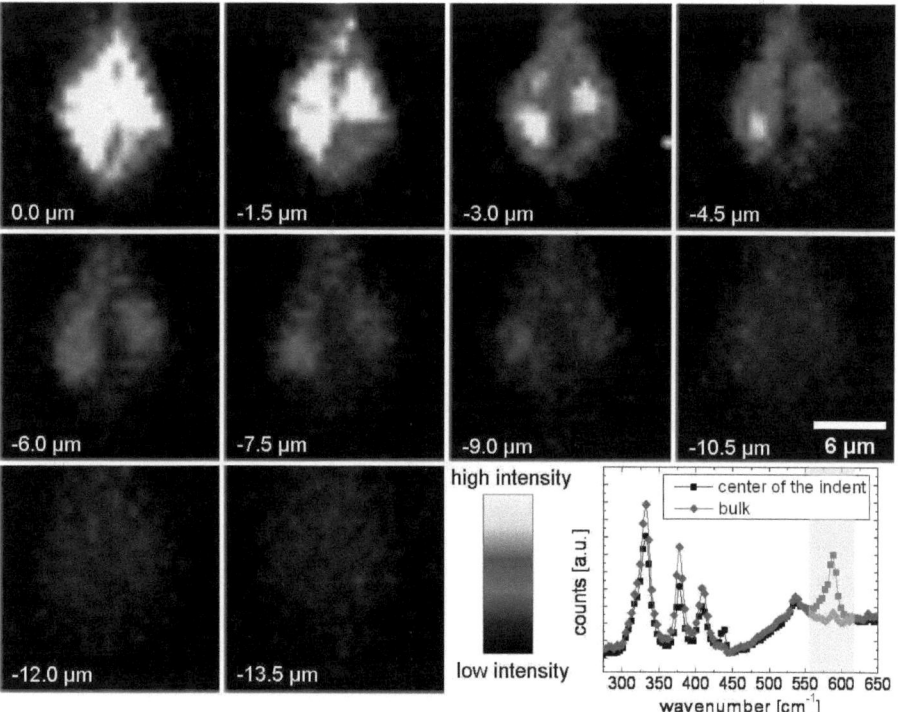

Figure 6.5. The black spectrum (squares) was measured in the center of the indent (see Figure 6.3 arrow I). The gray spectrum originates from a pristine area of the mapping (see Figure 6.3 arrow II). Intensity of the peak at around 583 cm^{-1} (marked in the spectra) in the vicinity of the indent. Starting from the surface, the scans were performed down to a depth of 13.5 μm with an interplanar distance of 1.5 μm.

From the stack of 2-D images 3-D images were calculated. This allowed visualizing the intensity distribution of the peak at 583 cm^{-1} in all directions (see Figure 6.6). As every color represents certain intensity the volume of the region where the peak appeared could be calculated. With a threshold set to 20% of the maximal intensity the volume calculated was 226 μm^3.

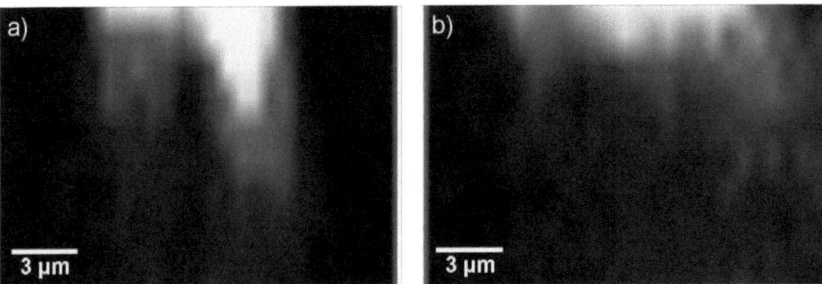

Figure 6.6. Intensity images from the peak at around 583 cm^{-1}. a) A cross section through the center of the indent along the z-axis. b) A cross section through the center of the indent parallel to the x-axis.

6.5 Discussion

Under hydrostatic conditions a peak shift directly correlates to the applied stress. Shifts to higher wavenumbers correspond to compressive stresses while a decreased peak position indicates tensile stress. Directly below the center of the indent, the residual stress is assumed to be hydrostatic. Manjon, Syassen and Lauck [23] measured a hydrostatic stress dependence of 4.33 cm^{-1}GPa^{-1} for the peak at 378 cm^{-1} and 5.2 cm^{-1}GPa^{-1} for the peak at 409 cm^{-1}. 3 μm directly below the center of the indent the peak at 378 cm^{-1} was increased by 1 cm^{-1} while the peak at 409 cm^{-1} was increased by 1.3 cm^{-1}. Using the peak shifts, the calculated stress was approximately -230 MPa to -250 MPa. The peaks at 378 cm^{-1} and 409 cm^{-1} are polarized perpendicular to each other. Therefore, two components of the stress tensor could be measured independently. The similarity of the stress values shows that the assumption of hydrostatic stress is reasonable.

In more complex stress states the direct correlation between peak shift and stress state is not valid. As the peaks at 378 cm^{-1} and 409 cm^{-1} originate from two different crystal orientation, two independent components of the stress tensor can be calculated. For accurate results one has to know the relation between strain, stress and Raman peak shifts, therefore the so-called phonon deformation potentials has to be calculated. In the region of the hydrostatic pressure one can calculate the strain tensor using the assumption that $\varepsilon_{xx} = \varepsilon_{yy}$ and the following equations which derive from Hooke's law:

$$\sigma_{xx} = (C_{11} + C_{12}) \cdot \varepsilon_{xx} + C_{13} \cdot \varepsilon_{cc},$$

$$\sigma_{cc} = 2 \cdot C_{13} \cdot \varepsilon_{xx} + C_{33} \cdot \varepsilon_{cc},$$

(6.2)

where $\sigma_{xx} = \sigma_{cc} = -250$ MPa and C_{ij} are stiffness constants with the following values: C_{11} = 195.4 GPa, C_{12} = 111.2 GPa, C_{13} = 92.5 GPa and C_{33} = 199.8 GPa [32]. The relation between the Raman peak shift and the strain tensor for ZnO is given by the following equations [33]:

$$\Delta\omega_{A1} = a_{A1} \cdot (\varepsilon_{xx} + \varepsilon_{yy}) + b_{A1} \cdot \varepsilon_{cc},$$

$$\Delta\omega_{E1} = a_{E1} \cdot (\varepsilon_{xx} + \varepsilon_{yy}) + b_{E1} \cdot \varepsilon_{cc} + c_{E1} \cdot \left[(\varepsilon_{xx} - \varepsilon_{yy})^2 + 4 \cdot \varepsilon_{xy}^2\right]^{1/2},$$

$$\Delta\omega_{E2} = a_{E2} \cdot (\varepsilon_{xx} + \varepsilon_{yy}) + b_{E2} \cdot \varepsilon_{cc} + c_{E2} \cdot \left[(\varepsilon_{xx} - \varepsilon_{yy})^2 + 4 \cdot \varepsilon_{xy}^2\right]^{1/2}.$$

(6.3)

$\Delta\omega_i$ is the peak shift and a_i, b_i and c_i are the phonon deformation potentials of the peaks at $i = A_1$ (378 cm^{-1}), E_1 (409 cm^{-1}) and E_2 (438 cm^{-1}). In the case of hydrostatic pressure the term $\left[(\varepsilon_{xx} - \varepsilon_{yy})^2 + 4 \cdot \varepsilon_{xy}^2\right]^{1/2}$ vanishes. Gruber et al.[22] calculated the phonon deformation potential for the E_2-mode at 438 cm^{-1} and obtained: a_{E2} = (-690 ± 110) cm^{-1} and b_{E2} = (-940 ± 260) cm^{-1}. Using these phonon deformation potentials, the strains calculated with equation (6.2) and the relations between the phonon deformation potentials and the strains given by Davydov et al.[34]

$$\varepsilon_{xx} = \frac{\frac{\Delta\omega_{A1}}{b_{A1}} - \frac{\Delta\omega_{Ei}}{b_{Ei}}}{2\cdot(\frac{a_{A1}}{b_{A1}} - \frac{a_{Ei}}{b_{Ei}})}, \text{ and } \varepsilon_{cc} = \frac{\frac{\Delta\omega_{A1}}{a_{A1}} - \frac{\Delta\omega_{Ei}}{a_{Ei}}}{\frac{b_{A1}}{a_{A1}} - \frac{b_{Ei}}{a_{Ei}}},$$ (6.4)

where $i = 1,2$ one can calculate the phonon deformation potentials for all the peaks. For the peak at A_1-mode at 378 cm^{-1} we obtained a_{A1} = -584 cm^{-1} and b_{A1} = -796 cm^{-1}. The phonon deformation potential of the E_1-peak at 409 cm^{-1} are a_{E1} = -568 cm^{-1} and b_{E1} = -1675 cm^{-1}. The term $[(\varepsilon_{xx} - \varepsilon_{yy})^2 + 4\cdot\varepsilon_{xy}^2]^{1/2}$ describes the splitting of the degenerated E_1- and E_2-modes. Significant splitting leads to a broadening of the peak width. Such a behavior was not observed in any measurement. Therefore, the phonon deformation potential c (see equation (6.3)) is very small and can be neglected.

On the surface the stress state can be assumed as biaxial with the out-of-plane stress component σ_{yy} = 0. Further, the main deformation systems are parallel to the c-axis consequently one can assume that the principle stress axes of the stress tensor correspond to the c- and x-axis, respectively. Therefore, the phonon deformation potentials and equation (6.2) allows calculating σ_{xx} and σ_{cc} at the surface of the crystal.

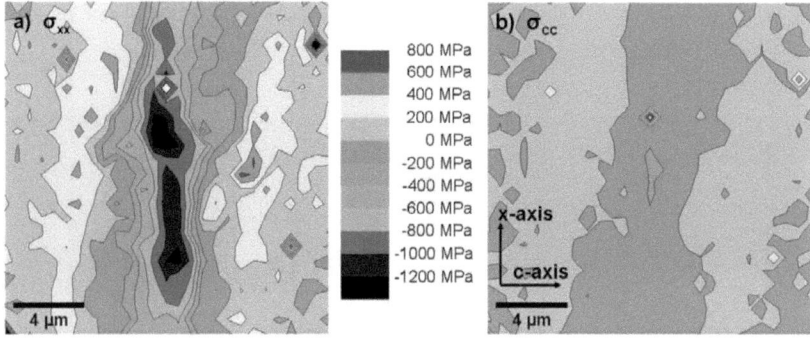

Figure 6.7. a) the σ_{xx}- and the b) σ_{cc}-component of a biaxial stresses field around the indent at the surface of the crystal.

Despite the fourfold symmetry of the Vickers indent the stress field at the surface (see Figure 6.7) shows a clear twofold symmetry. The σ_{xx}-component illustrated in Figure 6.7 a) shows a band of compressive stresses parallel to the x-axis and through the center of the indent. The compressive stresses are up to -1200 MPa. Along the region of compressive stress two bands of tensile stresses with values up to 800 MPa are detected. The σ_{cc}-component shown in Figure 6.7 b) shows a much less pronounced pattern. The residual stresses in the direction of the c-axis are much lower than in direction of the x-axis. But also here the same twofold symmetry is visible. In the center of the indent compressive stresses of about -400 MPa were calculated. To correlate these results with the ZnO-microstructure around the indent SEM and CL measurements were performed. The SEM image of the indent (Figure 6.8 a)) shows a crack parallel to the basal plane. The CL image in Figure 6.8 b) reveals two interesting facts. Firstly, the induced defects and dislocations, which appear as black, show a twofold symmetry. Dislocations extend mainly parallel to the basal plane. Secondly, the perturbed zone is much larger then the indent itself and the induced defects propagate up to 15-20 μm into the material. Both findings are in agreement with literature [35, 36]. The results of the CL measurement show that the residual stress field of the indentation is not determined by the shape of the indent but by the mechanism of plastic deformation. The CL image also gives an explanation why the amount of residual stresses differs so strongly between the c- and the *x*-axis. The majority of the dislocations prolong along the *x*-axis. Every dislocation causes a distortion which is much larger parallel to the *x*-axis then parallel to the *c*-axis.

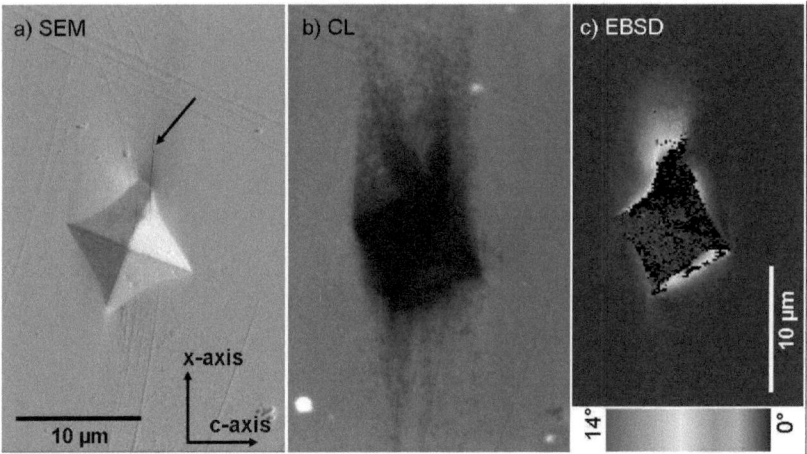

Figure 6.8. a) SEM image of the indent. The arrow highlights a crack parallel to the basal plane. b) CL image of the same indent. The defects and dislocations appear as black. c) EBSD image of the rotation of the out-of-plane orientation of the crystal around the indent. The local crystal orientation is rotated maximally 14° in comparison to the pristine ZnO crystal orientation.

The CL image explains the two fold symmetry of the residual stresses but it can not explain the pattern of the depth-scan along the x-axis (Figure 6.4 a)). Indentation studies performed on the basal plane of ZnO [12, 37-39] demonstrated that defects created by loading propagate in well-defined directions. First slip systems along the [1$\bar{2}$10], [2$\bar{1}$$\bar{1}$0] and [11$\bar{2}$0] directions are activated and extend deep into the bulk material. With increasing load, features which are offset by ~ 30° appear along the [$\bar{1}$010], [0$\bar{1}$10] and [1$\bar{1}$00] directions. This leads to a rosette structure with a hexagonal symmetry. All the deformation systems are parallel to the basal plane. A schematic illustration of the structure is presented in Figure 6.9. In Figure 6.4 a) the angle between band II and band III as well as the angle between band III and V is approximately 60° while the angle between I-II and between III-IV is in the order of 30°. The CL image and the 2-D Raman scans showed the correlation between defect structures and residual stresses. Thus, we assume that the bands measured in Figure 6.4 a) correspond to defect structures either along a [1 $\bar{2}$ 10] or along a [1 $\bar{1}$ 00] direction. The depth scan along the c-axis (Figure 6.4 b)) supports our assumption. The peak is increased only directly below the indent and no other bands are visible which would correspond to deformation directions not parallel to the basal plane.

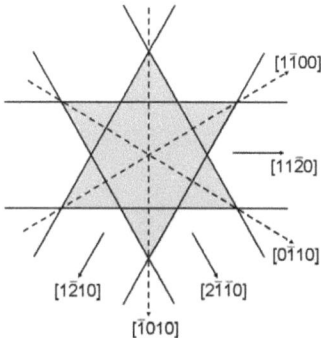

Figure 6.9. Schematic illustration of preferred directions for the propagation of dislocation caused by indentation on a ZnO basal plane [12, 35, 37].

Two explanations are possible for the increased peak intensity at 583 cm^{-1} which is shown in Figure 6.5. First, the indentation could have induced a local rotation of the crystal orientation. Secondly, oxygen vacancies as well as an excess of Zn at particular lattice sites or at interstitial positions can lead to increased peak intensity [26]. The orientation of the crystal around the indent was mapped by means of EBSD. The result illustrated in Figure 6.8 c) shows that the local out-of-plane orientation was rotated maximally 14° with respect to the pristine crystal orientation. The largest crystal rotations were observed at the pile-up and the borders of the indent while the crystal orientation of the indent was highly comparable to the pristine orientation. Regarding the intensity of the peak at 583 cm^{-1}, the rotation of the crystal orientation may play a role at the border of the indent but in the center of the indent the EBSD measurement shows that rotation can be excluded as the main cause for the increased intensity of the peak. Therefore, the increased intensity can be correlated to defect density.

Comparing Figure 6.5 and Figure 6.8 b) shows that only the center of the indent and therefore the region with the highest defect density causes an increase of the peak at 583 cm^{-1}.

6.6 Conclusion

The prism plane of a ZnO single crystal was indented with a Vickers microindenter. The indented was examined by confocal Raman microscopy, CL and EBSD and TEM. As ZnO is transparent it was possible to measure Raman maps in 3-D. The Raman measurements showed that the hydrostatic residual pressure 3 μm below the center of the indent are -230 to -250 MPa. The biaxial stress field on the surface of the indent showed a twofold symmetry. This can be explained by the deformation mechanism of ZnO. Dislocations prolongate mainly along the basal plane. σ_{xx} showed values between -1200 and 800 MPa while σ_{cc} showed stresses between -400 and 200 MPa. The distortion caused by the dislocations is much stronger along the x-axis then along the c-axis and is the reason for the big differences between the values of σ_{xx} and σ_{cc}. Raman depth-scans along the x- and the c-axis showed different behavior. The authors believe that the bands of increased peak position which can be seen in the depth-scan along the x-axis correspond to certain crystal orientations. In the center of the indent and directly below the center the Raman peak at 583 cm^{-1} was significantly increased. The increased intensity corresponds to oxygen vacancies and/or Zn interstitials. As the Raman maps were performed in 3-D it was possible to calculate the volume of the increased peak intensity which corresponds to the volume of high dislocation density and was 226 μm^3.

6.7 Acknowledgement

The authors thank Flavio Mornaghini for the preparation of the FIB lamella as well as for the TEM analysis. Matteo Seita and Karsten Kunze are acknowledged for their EBSD and CL measurements. The authors acknowledge support of the Electron Microscopy Centre of the Swiss Institute of Technology (EMEZ). This work was supported by the ETH Research Grant TH -39/05-1.

6.8 References

1. M. Ohtsu, K. Kobayashi, T. Kawazoe, S. Sangu, and T. Yatsui, *Nanophotonics: Design, fabrication, and operation of nanometric devices using optical near fields*. Ieee Journal of Selected Topics in Quantum Electronics, 2002. **8**(4): p. 839-862.
2. Z.L. Wang, *Zinc oxide nanostructures: growth, properties and applications*. Journal of Physics-Condensed Matter, 2004. **16**(25): p. DOI 10.1088/0953-8984/16/25/R01|PII S0953-8984(04)58969-5.
3. D.C. Look, *Recent advances in ZnO materials and devices*. Materials Science and Engineering B-Solid State Materials for Advanced Technology, 2001. **80**(1-3): p. 383-387.
4. U. Ozgur, Y.I. Alivov, C. Liu, A. Teke, M.A. Reshchikov, S. Dogan, V. Avrutin, S.J. Cho, and H. Morkoc, *A comprehensive review of ZnO materials and devices*. Journal of Applied Physics, 2005. **98**.
5. S.J. Pearton, D.P. Norton, L.C. Tien, and J. Guo, *Modeling and Fabrication of ZnO Nanowire Transistors*. Ieee Transactions on Electron Devices, 2008. **55**(11): p. 3012-3019.
6. M.H. Huang, S. Mao, H. Feick, H.Q. Yan, Y.Y. Wu, H. Kind, E. Weber, R. Russo, and P.D. Yang, *Room-temperature ultraviolet nanowire nanolasers*. Science, 2001. **292**(5523): p. 1897-1899.
7. M.Q. Wang, Y.Q. Lian, and X.G. Wang, *PPV/PVA/ZnO nanocomposite prepared by complex precursor method and its photovoltaic application*. Current Applied Physics, 2009. **9**(1): p. 189-194.
8. J.S. Ahearn, J.J. Mills, and A.R.C. Westwood, *Effect of electrolyte pH and bias voltage on hardness of (0001) ZnO surface*. Journal of Applied Physics, 1978. **49**(1): p. 96-102.
9. H. Ruf and A.G. Evans, *Toughening by monoclinic zirconia*. Journal of The American Ceramic Society, 1983. **66**(5): p. 328-332.

10. S.O. Kucheyev, J.E. Bradby, J.S. Williams, C. Jagadish, and M.V. Swain, *Mechanical deformation of single-crystal ZnO*. Applied Physics Letters, 2002. **80**(6): p. 956-958.

11. R. Navamathavan, S.J. Park, J.H. Hahn, and C.K. Choi, *Nanoindentation 'pop-in' phenomenon in epitaxial ZnO thin films on sapphire substrates*. Materials Characterization, 2008. **59**(4): p. 359-364.

12. J.E. Bradby, S.O. Kucheyev, J.S. Williams, C. Jagadish, M.V. Swain, P. Munroe, and M.R. Phillips, *Contact-induced defect propagation in ZnO*. Applied Physics Letters, 2002. **80**(24): p. 4537-4539.

13. R. Ghosh, D. Basak, and S. Fujihara, *Effect of substrate-induced strain on the structural, electrical, and optical properties of polycrystalline ZnO thin films*. Journal of Applied Physics, 2004. **96**(5): p. 2689-2692.

14. Y. Ohno, H. Koizumi, T. Taishi, I. Yonenaga, K. Fujii, H. Goto, and T. Yao, *Optical properties of dislocations in wurtzite ZnO single crystals introduced at elevated temperatures*. Journal of Applied Physics, 2008. **104**(7).

15. Y. Gogotsi, C. Baek, and F. Kirscht, *Raman microspectroscopy study of processing-induced phase transformations and residual stress in silicon*. Semiconductor Science And Technology, 1999. **14**(10): p. 936-944.

16. Z. Iqbal and S. Veprek, *Raman-scattering form hydrogentaed microcrystalline and amorphous-silicon*. Journal of Physics C-Solid State Physics, 1982. **15**(2): p. 377-392.

17. I. DeWolf, *Micro-Raman spectroscopy to study local mechanical stress in silicon integrated circuits*. Semiconductor Science and Technology, 1996. **11**(2): p. 139-154.

18. R. Ossikovski, Q. Nguyen, G. Picardi, J. Schreiber, and P. Morin, *Theory and experiment of large numerical aperture objective Raman microscopy: application to the stress-tensor determination in strained cubic materials*. Journal Of Raman Spectroscopy, 2008. **39**(5): p. 661-672.

19. T. Wermelinger and R. Spolenak, *Correlating Raman peak shifts with phase transformation and defect densities: a comprehensive TEM and Raman study on*

silicon. Journal of Raman Spectroscopy, 2009. Published Online: Feb 27 2009; 6:05AM DOI: 10.1002/jrs.2181.

20. T. Wermelinger, C. Borgia, C. Solenthaler, and R. Spolenak, *3-D Raman spectroscopy measurements of the symmetry of residual stress fields in plastically deformed sapphire crystals.* Acta Materialia, 2007. **55**(14): p. 4657-4665.

21. S.S. Mitra, O. Brafman, W.B. Daniels, and R.K. Crawford, *Pressure-Induced Phonon Frequency Shifts Measured By Raman Scattering.* Physical Review, 1969. **186**(3): p. 942-&.

22. T. Gruber, G.M. Prinz, C. Kirchner, R. Kling, F. Reuss, W. Limmer, and A. Waag, *Influences of biaxial strains on the vibrational and exciton energies in ZnO.* Journal of Applied Physics, 2004. **96**(1): p. 289-293.

23. F.J. Manjon, K. Syassen, and R. Lauck, *Effect of pressure on phonon modes in wurtzite zinc oxide.* High Pressure Research, 2002. **22**(2): p. 299-304.

24. S.K. Sharma and G.J. Exarhos, *Raman spectroscopic investigation of ZnO and doped ZnO films, nanoparticles and bulk material at ambient and high pressures.* Solid State Phenomena, 1997. **55**: p. 32-37.

25. F. Decremps, J. Pellicer-Porres, A.M. Saitta, J.C. Chervin, and A. Polian, *High-pressure Raman spectroscopy study of wurtzite ZnO.* Physical Review B, 2002. **65**(9).

26. G.J. Exarhos and S.K. Sharma, *Influence of processing variables on the structure and properties of ZnO films.* Thin Solid Films, 1995. **270**(1-2): p. 27-32.

27. C.F. Windisch, G.J. Exarhos, C.H. Yao, and L.Q. Wang, *Raman study of the influence of hydrogen on defects in ZnO.* Journal of Applied Physics, 2007. **101**(12).

28. X. Wang, Q.Q. Li, Z.B. Liu, J. Zhang, Z.F. Liu, and R.M. Wang, *Low-temperature growth and properties of ZnO nanowires.* Applied Physics Letters, 2004. **84**(24): p. 4941-4943.

29. R. Loudon, *Raman Effect In Crystals.* Advances In Physics, 1964. **13**(52): p. 423-&.

30. W. Rasband, *Image J 1.38x.* 2006, National Institutes of Health, USA: Bethesda, Maryland.

31. T.C. Damen, S.P.S. Porto, and B. Tell, *Raman effect in zinc oxide*. Physical Review, 1966. **142**(2): p. 570-&.
32. I.R. Shein, V.S. Kiiko, Y.N. Makurin, M.A. Gorbunova, and A.L. Ivanovskii, *Elastic parameters of single-crystal and polycrystalline wurtzite-like oxides BeO and ZnO: Ab initio calculations*. Physics of the Solid State, 2007. **49**(6): p. 1067-1073.
33. R.J. Briggs and A.K. Ramdas, *Piezospectroscopic study of Raman-spectrum of cadmium-sulfide*. Physical Review B, 1976. **13**(12): p. 5518-5529.
34. V.Y. Davydov, N.S. Averkiev, I.N. Goncharuk, D.K. Nelson, I.P. Nikitina, A.S. Polkovnikov, A.N. Smirnov, M.A. Jacobsen, and O.K. Semchinova, *Raman and photoluminescence studies of biaxial strain in GaN epitaxial layers grown on 6H-SiC*. Journal of Applied Physics, 1997. **82**(10): p. 5097-5102.
35. S. Basu and M.W. Barsoum, *Deformation micromechanisms of ZnO single crystals as determined from spherical nanoindentation stress-strain curves*. Journal of Materials Research, 2007. **22**: p. 2470-2477.
36. V.A. Coleman, J.E. Bradby, C. Jagadish, and M.R. Phillips, *A comparison of the mechanical properties and the impact of contact induced damage in a- and c-axis ZnO single crystals*. Zinc Oxide and Related Materials, 2007. **957**: p. 213-218.
37. Z. Takkouk, N. Brihi, K. Guergouri, and Y. Marfaing, *Cathodoluminescence study of plastically deformed bulk ZnO single crystal*. Physica B-Condensed Matter, 2005. **366**(1-4): p. 185-191.
38. V.A. Coleman, J.E. Bradby, C. Jagadish, and M.R. Phillips, *Observation of enhanced defect emission and excitonic quenching from spherically indented ZnO*. applied physics letters, 2006. **89**(8).
39. M.J. Klopfstein and D.A. Lucca, *Observation of nanoindentation rosettes on {0001}ZnO using scanning Kelvin probe microscopy*. applied physics letters, 2005. **87**(13).

7 Conclusion and Outlook

7.1 Conclusion

The aim of the work was to analyze mechanical stresses as well as microstructural changes of different materials in small dimensions by means of Raman microscopy. The results of the mainly experimental work are presented in 5 chapters. The particular chapters are dedicated to different aspects of stresses and microstructural changes analyzed in one-, two- and three dimensions. From the results it is possible to draw some general conclusions which can be divided into two main sections. The first section concentrates on the method Raman microscopy itself while the second section focuses mainly on mechanical stresses and relating materials aspects.

7.1.1 Raman Microscopy

Raman microscopy has proven to be an accurate method for measuring strain and stresses. In uniaxial compressive experiments performed on silicon micro-pillars the stresses derived from Raman microscopy and a load cell show were in very good agreement since the discrepancy between both methods throughout the test up to a load of about 5.1 GPa is less than 3%. In a highly Raman active material such as silicon the method has a strain resolution of 10^{-4} which allows to detect stresses as small as ~20 MPa.

Two-dimensional stress mappings of silicon thin films proved that the method is also able to measure stresses indirectly in metal thin films. Thermal stresses were applied by heating up the silicon-silicon nitride-aluminum multilayer structure up to 200°C. The temperature-induced stresses were measured by means of Raman microscopy and compared to theoretical considerations. At 200°C with both methods in the aluminum thin film compressive stresses of about -100 to -150 MPa were calculated. This approach still has a lot of potential for improvements. For example, replacing the polycrystalline silicon thin film with a single-crystal silicon film would increase the accuracy of the method. It would also allow analyzing inhomogeneous stresses in thin films.

Due to the confocality of the Raman microscope it was feasible to map stresses in three dimensions. The key requirement is to have transparent and Raman active samples. As the spatial resolution decreases with increasing penetration depth into the material the method is limited to some tens of micrometers. The number of independent components of the strain or stress tensor which are available by Raman microscopy depends on the material. 3-D stress states are in general of complicated nature. If only one or two components are measurable directly, one has to assume certain stress states to be able to quantify the stresses. Nevertheless, the measurements give a detailed qualitative impression about the appearance of stresses.

Besides performing Raman measurements the laser was used for inducing grain growth in amorphous silicon thin films. The produced grains in chapter 3 had an average diameter of 35 ± 17 nm. Other experiments showed that it was possible to produce grains with a diameter of up to 1 µm. While laser induced grain growth is well known for large scale applications, the small diameter of the laser allows selective grain growth in highly specific areas and producing crystalline structures with a size in the micrometer range. Since the crystalline silicon possesses material properties different than that of the amorphous silicon such as anisotropic etching rates for its crystal planes or electrical conductivity this could be a useful tool to produce freestanding structures or electrical conductors.

7.1.2 Materials Aspects

The micro silicon pillars had an average compressive failure strength measured in the middle of the micro-pillars of -5.1 GPa. In all the performed measurements the stress-strain-correlation of the pillars was linear. In a small number of measurements a shoulder on the low energy side of the silicon Raman peak was detected. Based on a TEM study, it was found that the main deformation mechanism of the pillars is brittle fracture. Dislocation nucleation and movement seems to play only a secondary role and to be triggered by interaction with cracks. Phase transformations could neither be observed during (by Raman spectroscopy) nor after the deformation (by TEM). Further, phase transformations require a high strain rate which was not given in the performed experiments. Hence, we strongly believe that the shoulders observed are due to cracks propagating through the laser spots.

One advantage of Raman microscopy is that it additionally contains information about the microstructure of the probed materials. This is shown in several chapters of the thesis but especially in chapter 4. A TEM lamella was extracted from the center of an indent placed on a silicon wafer. As silicon is highly Raman active it was possible to map the lamella by means of Raman microscopy. The results illustrate how the peak width highly correlates with the defect density while the peak position is mainly altered by implanted Ga^+-ions which were needed for cutting out the lamella by means of FIB. The shifted Raman peak can be explained by the phonon confinement model (PCM). The implanted ions cause damage in the single crystalline silicon. Therefore, one assumes that phonons are not infinite as in the single crystalline case anymore but restricted by grain boundaries. The restriction causes a relaxation of the phonon wave vector. The smaller the crystallite size the higher is the peak shift. This shows how the preparation of samples can influence the results.

The 3-D studies of sapphire and of ZnO illustrated very clearly that the residual stress fields is primarily influenced by the plastic deformation mechanism and not by the geometry of the indent. In both cases a Vickers indent was used to plastically deform the basal plane of a sapphire single crystal and the prism plane of a zinc oxide single crystal. Instead of the fourfold of the symmetry of Vickers indent the residual stress fields around the indent showed a threefold symmetry in the case of sapphire and a twofold symmetry in the case of zinc oxide, respectively. In ZnO the indent was placed on the prism plane to be able to detect Raman signal from two phonon vibrations which are perpendicular to each other. This allows quantifying biaxial stress on the surface of the crystal. In ZnO the dislocations prolong mainly along the basal plane which leads to the already mentioned twofold symmetry of the residual stresses. σ_{xx} showed values between -1200 and +800 MPa while σ_{cc} showed stresses between -400 and +200 MPa. The distortion caused by the dislocations is much stronger along the x-axis then along the c-axis and is the reason for the big differences between the values of σ_{xx} and σ_{cc}.

In both 3-D studies changes of the microstructure could be observed. In the case of sapphire, the Raman signal from directly below the indent is significantly altered. This most probably a result of a stress induced phase transformation. In ZnO no phase transformation is visible. Nevertheless, directly below the indent the intensity of the peak at 583 cm^{-1} is strongly increasing. The increased intensity possibly corresponds to oxygen vacancies and/or Zn interstitials. As the Raman maps were performed in 3-D it was possible to calculate the volume of the increased peak intensity which corresponds to the volume of high dislocation density and was 226 μm^3. The theoretical limit is the volume of one single voxel which has a volume of approximately 1 μm^3.

7.2 Outlook

The results of this work showed that Raman microscopy is an appropriate method for examining different stress and microstructural related questions. One of the major tasks of future developments for all applications of Raman microscopy is to increase the spatial resolution. One potential way is the use of UV lasers. This approach has the disadvantage that the increased energy of the incoming light also increases problems regarding fluorescence. Other approaches are combining either scanning near-field optical microscopy (SNOM)-technology or surface enhanced Raman scattering (SERS)-technology with Raman microscopy. Both approaches have the potential to significantly increase the lateral resolution. Some preliminary studies, using the SERS-effect for enhancing the spatial resolution, were accomplished in the scope of this thesis. The results of the studies were somehow frustrating. Different materials such as gold, silver and platinum as well as different grain sizes were under examination. The measured

SERS effect was very weak if it was detectable at all. Reasons are the (for SERS) disadvantageous backscattering set-up of the microscope as well as a not ideal particle size. Special arrangements of nanoparticles might increase the SERS-effect. Although the results were not very promising up to now, SERS is a too powerful method to be neglected. For further studies it would be of greatest interest to cooperate with groups which have much more experiences in the fields of SERS concerning simulations and experimental work.

As already mentioned in the conclusion, it is possible with silicon-silicon nitride-metal multilayer structures to measure stresses in the metal thin film by means of Raman spectroscopy. In the presented case study a poly-silicon thin film was used. Replacing the poly-crystalline film with a single-crystal film would increase the lateral resolution as well as the accuracy of the method. The experimental set-up is mainly suitable for measuring residual, thermal stresses and stresses applied by means of bulge-tests. Due to the fast data acquisition and the good lateral resolution of the Raman microscope it would be possible to perform more general studies in terms of different metals, film thicknesses and grain sizes. Another material which has the potential to be used as a strain gage is mica. The advantage of that material is the flatness of the surface and that it is single crystalline. As it is transparent it would be possible to focus directly on the mica-metal interface.

One of the main conclusions of the thesis is that 3-D residual stresses in the vicinity of a micro-indent are mainly influenced by the deformation mechanism of the material rather than by the geometrical shape of the indent. A very interesting experiment would be to see how the stress field behaves during *in situ* micro-indentation. Another fascinating aspect could be how the residual stress field appears for larger indents or indents with special geometry (high length/width-ratio etc.). For quantifying stresses exactly one has to know the phonon deformation potentials of the materials. These values are only known for few materials and often the values stated in different publications show large deviations. Therefore it would be of great interest to define the phonon deformation potentials for different materials using XRD, Raman and other methods.

We showed that the peak positions of polymers also depend on the applied stresses. Moreover, we could illustrate how the peak shift is depending on the orientation of the polymerchains. The great advantage of polymers the large amount of strain which can be applied. The measurements were disturbed by slip between the polymer samples and the clamps of the tensile machine. Further investigations have to find a sample set-up which minimizes the problem of slip.

Due to the high power and the very small focus spot of the laser it is possible to heat up samples to a very high temperature. On one side, this can be a potential problem for experiments with polymer as the laser might destroy the sample. On the other side, the high energy density can be used for induce grain growth in amorphous silicon thin films. Some rough estimation showed that the silicon thin films are locally heated up to more than 1200°C. The method could be used to produce crystalline structures with a size in the micrometer range. As the crystalline silicon has other material properties than the amorphous silicon such as etching rates or electrical conductivity this could be a useful tool to produce freestanding structures or electrical conductors. The method is not only limited to silicon thin films but might be also applied to other Raman active materials. One could also think of locally heating metals indirect using an experimental set-up such as used in chapter 3. For example sputtering a steel thin film with a martensitic crystal structure could be locally heated up above the austenitic temperature. One might obtains a non-magnetic thin film which features defined areas of austenitic magnetic structures.

I want morebooks!

Buy your books fast and straightforward online - at one of the world's fastest growing online book stores! Environmentally sound due to Print-on-Demand technologies.

Buy your books online at
www.get-morebooks.com

Kaufen Sie Ihre Bücher schnell und unkompliziert online – auf einer der am schnellsten wachsenden Buchhandelsplattformen weltweit! Dank Print-On-Demand umwelt- und ressourcenschonend produziert.

Bücher schneller online kaufen
www.morebooks.de

OmniScriptum Marketing DEU GmbH
Heinrich-Böcking-Str. 6-8
D - 66121 Saarbrücken
Telefax: +49 681 93 81 567-9

info@omniscriptum.com
www.omniscriptum.com

Printed by Books on Demand GmbH, Norderstedt / Germany